CELEBRATING

50 YEARS

Texas A&M University Press

publishing since 1974

Migration Mysteries

W. L. Moody Jr. Natural History Series

Migration Mysteries

Adventures, Disasters, and Epiphanies in a Life with Birds

John H. Rappole

TEXAS A&M UNIVERSITY PRESS
COLLEGE STATION

Library of Congress Cataloging-in-Publication Data

Names: Rappole, John H., author.
Title: Migration mysteries: adventures, disasters, and epiphanies in a life with birds /
    John H. Rappole.
Other titles: W.L. Moody Jr. natural history series; no. 42.
Description: First edition. | College Station: Texas A&M University Press, [2024] |
    Series: W.L. Moody Jr. natural history series; [no. 42] | Includes bibliographical
    references and index.
Identifiers: LCCN 2023043538 (print) | LCCN 2023043539 (ebook) |
    ISBN 9781648431838 (cloth) | ISBN 9781648431845 (ebook)
Subjects: LCSH: Rappole, John H. | Ornithologists—United States—Biography. |
    Birds—Migration. | Ornithology—Fieldwork. | LCGFT: Autobiographies.
Classification: LCC QL31.R2 R36 2024 (print) | LCC QL31.R2 (ebook) |
    DDC 598.092 [B]—dc23/eng/20231102
LC record available at https://lccn.loc.gov/2023043538
LC ebook record available at https://lccn.loc.gov/2023043539

For Bonnie Elizabeth, again

Oh, you were the best of all my days.
—FRANK O'HARA, *ANIMALS*

CONTENTS

PREFACE

And again he sent forth the dove out of the ark;
And the dove came in to him in the evening; and, lo, in her mouth was an olive
leaf pluckt off: so Noah knew that the waters were abated from off the earth.
And he stayed yet other seven days; and sent forth the dove;
which returned not again unto him any more.

—Genesis 8:10–12

Migrants are doomed.

Such is the current understanding of bird migration as expressed in the popular literature as well as by many scientists. The reason given is migrants' need for a multiplicity of habitats, often hundreds or thousands of miles apart, during different portions of the annual cycle. This dependency supposedly makes them highly vulnerable to a multitude of threats to their populations. Several recently published works on migration are blaring this alarm. The authors, each striving to be the Rachel Carson of the twenty-first century, stress that with all the human-caused changes occurring, migrants are like "canaries in the coal mine," their populations sensitive to the slightest change in the ecological atmosphere.

Exemplary of this genre is an account I read this past spring (April 2021) written by a skilled professional employing many of the tools in the creative writing bag of tricks: scene setting, engaging and alarming storytelling heightened by hyperbole, character development, extended metaphor, and of course, an emergency—*Migrants Are in Dire Straits Due to Climate Change, Signaling the End of the World as We Know It!*

The book focuses on two main themes:

> **Discovery**—using the author's personal experiences and contacts to describe how new technology has revealed (purportedly) previously unknown aspects of the migrant world

Disaster—enumerating how these discoveries have exposed the difficulties in which many migrant species find themselves and what these mean for us

This gentleman and the other Cassandras are wrong. Migrants are fine, and most will be around long after our own interesting species has joined the throngs of the departed. Explaining why we need not worry about migrants will require a lot of discussion, which is what this book is all about. I began my investigations of their biology half a century ago, and it is through the results of that research that I have come to understand that appalling prognostications concerning migrants are not fact based. Given the slightest chance, migrants will find what they need to survive. Migration is not about vulnerability; it is about flexibility. Noah's dove is a far better metaphor than gasping canaries.

But why is the common view of migrant needs and vulnerability wrong? The reason is that our current theories for the phenomenon are based on an ancient misunderstanding of the relationship between migration and weather.

Figure P-1. Mallards and snow, Jamestown, New York, November 15, 2022.

"Like cranes in clamorous lines before the face of heaven, beating away from winter's gloom and storms over the streams of Ocean, hoarsely calling." This quote from the *Iliad* serves as a clear statement of that relationship, implying that without change in weather, there would be no migration, an idea held for millennia that continues to the present day among both the public at large and the majority of the scientific community. In the context of the origin and evolution of bird migration, it is known as the northern home theory.

Like most everybody else, the perspective from which I began my quest was that provided by this theory, which holds that migratory birds originated on their breeding grounds as resident (nonmigratory), sedentary members of these communities. Climate change in these northern places occurring over millions of years favored the evolution of the characteristics that enable birds to leave for more equable climes when the weather is bad and return months later when the weather has ameliorated.

A series of unplanned events (accidents, bonanzas, catastrophes) in my life gave me the opportunity to take a closer look at this existing paradigm for migration. The tools used were those given to us by the great philosophers of science. In broad terms, these can be summarized as follows:

Knowledge—Existing knowledge on the topic of your interest is always the starting point. You have to know what others have discovered and the main theories explaining how your discipline works.

Investigation—You have to have the opportunity to collect data relevant to the principal questions in your field.

Skepticism—If data don't fit an existing explanation, whether that of someone else or your own, you have to accept that fact and consider alternative explanations.

Subjection to Logic—This is perhaps the most important tool in the scientist's bag and the most difficult to master: "If A is true, and B is true, then A + B = C must also be true." And, of course, if something you discover causes you to question the validity of A or B, then C also is on shaky ground. Mathematics is logic on steroids, allowing you to become increasingly sophisticated in the kinds of questions you can ask. Nevertheless, the sad truth is that most people hate logic. Belief will almost always trump logical conclusions.

Questioning of Assumptions—If people in general hate logic more than anything else, one could argue that the questioning of assumptions is most hated by scientists. More often than not, it is the questioning and critical examination of the assumptions involved in the construction of a theory that end in its destruction. Assumptions constitute the foundation of nearly all great theories simply because there are certain kinds of data that are difficult to collect, at least at the time that the assumptions are made. The creation of a new theory almost always requires a leap of faith, which is where the "unknown unknowns" often sneak in undetected.

My research has led me to believe that weather is *not* the ultimate cause of bird migration. Dispersal is. That is, young birds forced to leave home by parents and neighbors in search of a seasonally suitable place to live and breed constitute the origin of migration. I call this idea the dispersal theory. It's a concept that has been around for a long time but is discounted by most migration theorists.

I have presented the scientific information supporting this hypothesis in more than a hundred scientific papers and several books. So what makes this particular book so necessary and special? Fair question. In response, I would argue that the approach is different. Here I present my pursuit of the truth regarding migration in ways similar to Caesar's method in his book *Gallic Wars* because I think the doing of the work will interest readers and help them build an understanding of the science. After all, you could summarize the results of Caesar's campaigns in a page or two. It's the planning, maneuvers, negotiations, and battles that keep us reading his account two thousand years after the events took place. The process is what demands and holds our interest, the humanity of the work as it happened. And that's what you will find here—how each new investigation built toward a new understanding of migration, starting with our first efforts to examine migrant lives in Texas chaparral and Mexican rainforests in the early 1970s and culminating with our work on range change in Texas in the early 2000s, which gave us the crucial final piece of the puzzle.

Thus, my focus in this saga is more on the journey and moments of discovery than the destination. Field biology—that is, the process of gathering information about how organisms live in the places where they live—is personal. You can't do it without going to those places, and you can't go to those places without things happening to you and the people who are with you. So that's what this book is about. It is an epic account of the struggles, adventures, calamities, mistakes, and disasters involved with figuring out how migration works—as lived by me, my family, and a cast of incredible people with whom I have shared the path. And one more thing—those awe-filled moments when you realize that something you just saw or learned changes everything.

Buckle up. It's a wild ride!

ACKNOWLEDGMENTS

I am a part of all that I have met.

—Alfred Lord Tennyson, *Ulysses*

Swans sing, according to Greek mythology, only once as they are about to pass out of existence. The term *swan song*, however, has come to include less-final departures and therefore seems appropriate as applied to this book, which I see as my final word on the subject of bird migration. There are critics who will scoff at this notion. "Hah! One can only hope!" some might say. But I really mean it this time, and that being the case, I will use these acknowledgments a bit differently from those found in my earlier writings on the subject wherein I tried to recognize every person who helped in any way with any aspect of the work. Here I step back to take in the larger picture, celebrating the contributions of those who made my seven decades in natural history studies possible.

My parents, Francesca Elizabeth Rappole (née Goodell) and Albertus Whitney Rappole, set me on this voyage and were the principal guides along it through my first decade—bird walks, nature readings, travels, museums, shows. My father hoped that I would see such experiences as the grounding for a life in applied science where you can actually earn a living. "Think of your bird studies as an avocation," he advised me, and although he was certainly disappointed when I deviated from medicine to follow the road less traveled (in truth, a rather obscure path through the weeds, as it turned out), there were no hysterics or threats—love and funding continued unabated.

Early instruction in the joys of discovery associated with bird-watching came from Clarence Beal. A landscaper by trade, he had been the great naturalist Roger Tory Peterson's closest friend when both were learning their birds as youths growing up in Jamestown, New York, my own hometown. He, along with my parents and several other enthusiasts, mostly teachers in the public school system, were founding members of

Figure A-1. Mom and Dad on a lake in the California High Sierras in 1943 just before Dad shipped out as a medical officer in the Army Air Corps for service in the South Pacific.

the Jamestown Audubon Society in the mid-1950s. Mr. Beal was endlessly patient with young aspirants to bird knowledge like me.

My three closest companions on birding trips in grade school and junior high were my sister, Rosemary; my girlfriend, Pam Westrom; and Craig Bohall. The shared enthusiasm for each new find was a beacon for us.

Roger Tory Peterson, like me, was a Jamestown boy. He came back every year in the 1950s and '60s to give programs, usually movies, of birds seen on his travels accompanied by his narration. Though we all met and spoke with him many times, his obvious skills as an artist, photographer, and raconteur made him a figure of inspiration rather than aspiration.

My first professional mentor was Dr. Robert Goodwin, my undergraduate advisor at Colgate. He had a PhD in mammalogy from Cornell, and it was he who convinced me that it was possible to support a family as a naturalist—namely, by following the same route as himself.

During the late '60s, Colgate had a remarkable program called the January studies period, a monthlong segment between the fall semester, which ended in December, and the spring semester, which began in February. Students could choose how to spend the period: one could either take a single, intensive course from among the many offered or design one's own as an independent study under the guidance of a sponsoring faculty member. During my freshman year, I took the first route, an intensive study of the book of Job offered by Dr. Berry of the Philosophy and Religion Department. Sophomore year, however, Dr. Goodwin encouraged my proposal to join my parents in St. Croix (US Virgin Islands) for a month to do an independent study of the birds of that Caribbean island, which I did. When I arrived in St. Croix in early January 1966 to pursue this study, the first person I tried to find was George Seaman. I knew from my literature review that Mr. Seaman had done much of the original work on the birds of the island, along with his lifelong friend Harry Beatty, and that he was still living on St. Croix as a member of the staff of the experiment station of the US Department of Agriculture (USDA) there. Mr. Seaman greeted me warmly when I tracked him down at his office and gave me a wealth of information and advice on where to go to look for what. This little investigation was my first occasion to do actual field research: reviewing the literature, going out and collecting data, and writing up the results in a publishable paper on the topic. It was this experience that led directly to my decision to abandon medicine and pursue a career as a research scientist. In addition, George helped me land my first actual research job—a summer position with the St. Croix USDA searching for the native foods and relatives of the tobacco hornworm (*Manduca sexta*), a sphinx moth whose larvae can cause significant damage to tobacco crops. Eventually, I was offered a full-time position there to continue that and related studies as an assistant to the entomologist in charge of the work, and I would have accepted it, but my draft board advised me that taking a leave of absence from Colgate would result in my immediate classification as 1-A, followed in short order by being drafted into the army.

I was a mediocre student at Colgate and no one's idea of a person with great potential as a field biologist—or much else, to be honest. Nevertheless, Dwain Warner, then a senior professor in one of the most prestigious graduate programs in the country in my field of ornithology, decided to take a chance on me. Back in those days, if a prominent faculty member wanted a particular student to join his research group, a call by him to the graduate office would suffice, regardless of grades or GRE (graduate record exam) scores. Although I was drafted out of graduate school shortly after joining Dwain's program in the fall of 1968, we managed a reboot in 1972, landing a substantial grant to support my PhD work. Dwain, for me, was the model of the major advisor. His door was always open, and he was perfectly happy to talk research ideas for as long as you wanted.

If he got a phone call during such discussions, he would simply ignore it: "I'm talking to you right now, and that's enough. When I'm done, I will be happy to talk to someone else, and they will receive the same courtesy." He had immense experience and skill not only as a field biologist but as an advisor and a fundraiser. His graduate students weren't starving, rudderless indigents, like many whom I have known.

Through Dwain's contacts and advice, I was able to obtain a grant from the Welder Wildlife Foundation of Sinton, Texas, to support my entire graduate research and career at the University of Minnesota. Dr. Clarence Cottam was the director at Welder, and it was he who reviewed our proposal and gave us our funding. I had the chance to meet him during our first visit to Welder in the spring of 1973, and he and my family met and dined together often during the short year we had before his death in April 1974. Our children loved him, as did the large stable of Welder fellows working at the refuge during that time, whom he often took to lunch at Moya's, a marvelous Tex-Mex place in nearby Refugio, Texas. He had a national reputation as a wildlife biologist and environmentalist, of course, but he was also a skilled administrator, guiding Welder into position as one of the premier institutions of its kind.

Mario Alberto Ramos Olmos was the Mexican student chosen by my advisor, Dwain Warner, to work with me on our Welder grants in 1973. That collaborative effort marked the beginning of a fifteen-year partnership conducting research on migratory birds in the Tuxtla Mountains of southern Veracruz, culminating in a sodden month of radio-tracking Wood Thrushes, on the ground and from the air, at the Tuxtla biology station of the Universidad Nacional Autónoma de México (UNAM). It was during that final piece of fieldwork together that we were able to prove why it is better to be a territory holder in rainforest than a wanderer in hedgerows (less chance of being eaten). Shortly thereafter, Mario accepted a fellowship designed specifically for him by Tom Lovejoy at the World Wildlife Fund (WWF), catapulting him from being a leader in Mexican field biology into the stratosphere of international conservation. He and his family moved from Mexico to a beautiful home in Bethesda outside DC, and in 1994 he joined the World Bank as senior environmental consultant, a position requiring tens of thousands of air miles to dozens of countries each year. He finally managed to work himself to death there in 2006.

I first met Dr. Eugene Siller Morton at the close of my first professional presentation at a national conference, the annual meeting of the American Ornithologists' Union (AOU) at the University of Oklahoma in October 1974. Gene was a Smithsonian research scientist at the time and among the world's leading experts on migrant ecology in the tropics. As such, he recognized immediately the importance of our discovery of territoriality in migrants wintering in the Veracruz rainforest and expressed interest in possible collaboration in future work. We corresponded over the next couple of years and eventually put together a National Science Foundation (NSF) proposal to use radio-tracking for studying the movements of wintering migrants in the rainforest of the Smithsonian Tropical Research Institute on Barro Colorado Island in the Panama Canal Zone. The grant was not funded, but this effort led to the most important professional relationship of my career. Gene is famous for using his considerable influence in tropical field biology to promote the careers of novice practitioners, and so it was with me. He got me the contract for the preparation of the first exhaustive literature review and summary on new-world migrants, helped me get funding for migrant research from the WWF, and led the search committee that recommended my hiring as research coordinator at the Smithsonian's Conservation and Research Center (now the Smithsonian Conservation Biology Institute). We also collaborated on research in Mexico and several publications.

Thomas J. Lovejoy III, the founder of the field of conservation biology and arguably its greatest practitioner, died on Christmas Day 2021. Like my friend Mario Ramos, he played at a different level in the ecological research and conservation community than most of the rest of us. While we were focusing on what the individual birds were doing, Tom was leading various countries of the world toward preserving their precious natural resources. His cred from a scientific perspective was impeccable: PhD in field biology from Yale followed by a postdoctoral fellowship with the most famous ecologist of his time, Princeton's Robert MacArthur, investigating bird communities of the Amazon basin. From his position as assistant secretary at the Smithsonian, he served as an advisor to presidents and Congress on such matters as swapping debt for preserving huge blocks of primary forest in countries throughout the Western Hemisphere. Earlier in his career, he had served as vice president of science at the WWF, and it was from this position that he gave critical help to Mario and me, providing tens of thousands of dollars in funding for our work on migratory birds in Mexico in the early and mid-1980s. Our research could not have been accomplished without his aid.

Unable to find a job as a research scientist after getting my PhD, I worked for two years as a high school biology teacher at a Minneapolis prep school (1976–78), then took a year as an unpaid postdoctoral fellow back at the Bell Museum of Natural History. Toward the end of that year, in May 1979, I applied for a top-level research position at the Georgia Museum of Natural History advertised by the museum's director,

Josh Laerm. To my surprise, he chose me for the position, basically on advice from one of his PhD students, Joe Meyers, whom I had met at one of the national meetings of the AOU. It was Josh who gave me my first real job in my profession. With funds available to states from section 6 of the Endangered Species Act, he put together a team to investigate Georgia's endangered species: Bud Freeman for fish, Lloyd Logan for mammals, Laurie Vitt for reptiles and amphibians, and myself for birds. Josh turned out to be an outstanding boss, and although our funding ran out after two years (thanks to Reagan's election), we got to do a lot of interesting stuff. He was also a tireless supporter of all of our subsequent careers. In addition to being an excellent manager and research scientist, he was a legendary classroom teacher, one of only two professors I have ever heard of to receive a standing ovation on completion of a lecture (my colleague at Texas A&I University Charlie Russell was the other). *My* students often had to prod one another when it was time to move on to the next class. Sadly, Josh died young of complications from abdominal surgery in 1997.

I was working on radio-tracking Wood Thrushes in early February at our La Peninsula site in the Santa Martas when Dick Vogt, famed herpetologist from the University of Mexico, walked up the path. I considered locating us a considerable feat, since we were six miles from the nearest road, but Dick, with his long experience as a field biologist, could track like Jim Chee (he passed away in Brazil in 2021). He was accompanied by two undergraduate students of Dwain's, John Gerwin and Kevin Winker. This incident was my first introduction to Kevin, who has been my collaborator on many Mexican projects since. He is both literally and figuratively a giant of a man. At six five, he has the physical stature as well as the strength and stamina of a horse, but unlike many such, he possesses an amazing intellect to go with his size. By the end of the year, he had begun graduate work with me, taking over the exceptionally demanding task of tracking the wanderings of Wood Thrushes across the crags and cataracts of the most remote parts of the Tuxtla Mountains. Now curator of birds at the University of Alaska's Museum of Natural History, Kevin played a large role in deciphering how migrants manage to survive the wintering period in tropical rainforests.

Sam Beasom, late director of the Caesar Kleberg Wildlife Research Institute (CKWRI) at Texas A&I University (now Texas A&M University–Kingsville), hired me to do research on White-winged Doves in the Rio Grande Valley after I had been terminated as curator of natural history at the Connor Museum in June 1983. Subsequently, he brought me on staff as a full-time research scientist for the institute, a position I held until leaving for the Smithsonian in 1989. Sam basically took me off the figurative

Figure A-2. Kevin Winker working on his field notes with Mario in Santa Marta, March 1985.

garbage heap and allowed me to rebuild my career. He died in 1995, far too young. I honor his memory as a leader of the wildlife profession and an outstanding person.

I first met George Van Nostrand Powell at the Smithsonian migratory bird conference in October 1977. Subsequent to the Front Royal meetings, I read his seminal papers on mixed-species flocks in tropical forest that included migrants with interest but had not had the chance to talk with him until we got together at an AOU meeting in the early 1980s. We found we had a shared interest in determining the amount of remaining tropical forest habitat and set about trying to find the collaborators and funders to make the project possible. Eventually, we were successful, conducting fieldwork in Costa Rica, Belize, and Mexico for nearly a decade. Throughout our collaboration, George was the consummate professional, whether in fundraising, fieldwork, or publication preparation.

When George Powell and I got money for our work using satellites and geographic information systems (GIS) along with ground surveys to identify migrant wintering habitat, I knew I needed a person with outstanding ability to help me. My dear friend and colleague Dick DeGraff, chief research wildlife biologist at the Northeastern Forest Experiment Station at the University of Massachusetts, recommended Dave King. Twenty-five years later after fieldwork on multiple projects together in Mexico, Guatemala, Honduras, Nicaragua, and Burma, this suggestion has been resoundingly confirmed. I have never met a person with greater field skills—and King can play the mandolin! Furthermore, despite being a couple of inches shorter and twenty or so pounds lighter than Kevin Winker, when it comes to dinner, he can out-eat him any day.

In August 1989, my family and I came to the Smithsonian's field site, the Conservation and Research Center, where I was to assume the role of "research coordinator," a quasi-administrative position—half management and half research, supposedly. At the time of my arrival, about 90 percent of the on-site research at the 3,600-acre facility in the Blue Ridge Mountains was being done by one person, Dr. Bill McShea. As a result, "coordination" quickly became "collaboration." Bill had come to the Smithsonian as a postdoctoral fellow a few years previously and had managed to find funds to conduct a study on the effects of high deer populations on the Blue Ridge oak forest ecosystem. I joined the project, adding a bird component, and we worked together on this and many other investigations in Virginia and Myanmar over the next twenty years. We raised a couple of million dollars to support our research over that time, which is not a surprising amount in our business. More noteworthy, perhaps, is that we cobbled these funds together from about thirty different sources, which is a heck of a lot of groveling no matter how you cut it. Bill is a superb field biologist and the best of men.

I met Barbara Helm at breakfast during a symposium held at the storied Radolfzell lab of the Max Planck Institute in 2001. She was then a graduate student—indeed, the last doctoral student of the preeminent authority on timing of migration, Professor Doctor Eberhard Gwinner ("Ebo" to friends). Barbara was familiar with most published work on the topic of migration, and she had some insightful questions regarding my ideas. I was impressed by both the depth of her knowledge of the literature and her ability to place her own findings into a meaningful context. Her subsequent publications with Dr. Gwinner and on her own after his death demonstrate a search for truth that seems to suffer little from the regional bias that affects most scientific investigations.

This balance may derive in part from her own history. She did a master's on Nietzsche at Berkeley in her youth, taking up her PhD work at Radolfzell only after raising her three kids. In my experience, the world of migration studies is divided into two schools—that of the United States and that of Europe (mostly German and British). This dichotomy likely stems from the tendency of US folks to publish (and read) mostly in US journals, while Europeans tend to stick to European journals. For whatever reason, Barbara suffered from no such parochialism. She was (and is) familiar with all the relevant literature, regardless of who did it or where it was published. In addition, she always brings an incisive and well-informed mind to its analysis. It was for these reasons that I invited

Figure A-3. Bon and me camping in the Wichita Mountains, Oklahoma, March 1970.

her to join me in my effort to summarize the entire field of migration ecology and evolution in the book *The Avian Migrant*, published by Columbia University Press in 2013. We struggled together through chapter 4 of that eleven-chapter book until she finally had enough of my flights of fancy and expressed her desire to drop out of the project. As a result, the book was less than what it might have been in terms of an incisive understanding of literature and new work from both sides of the Atlantic. Nevertheless, my association with her certainly made the effort superior to what it would have been without her help.

Bonnie Elizabeth Carlson and I met in Miss Barber's French class when we were both fifteen. That Christmas, I invited her to the "Fish Under the Sea" dance (as our kids call it), and here we are, over half a century later. During our early years together, as will be readily apparent in the first chapters of this book, Bon shared in my fieldwork as well as running the household. Later, she was left to earn money (as a registered nurse) and raise the increasingly obstreperous progeny on her own for long periods while I went off to buckle my swash in various foreign parts. Throughout, however, she has been my indispensable partner. It was Bonnie who, on reading an earlier version of this manuscript, sighed and said she wished there was some way to convey the moment of awe and elation that accompanies the realization that a single observation has fundamentally shifted the paradigm. Her comment made me realize that regardless of whatever else I might wish to accomplish with his book, giving some sense of that feeling must be the ultimate goal.

Migration Mysteries

RETURN FROM THE WARS

"There was a ship," quoth he.

—SAMUEL TAYLOR COLERIDGE, *THE RIME OF THE ANCIENT MARINER*

My feelings on mustering out of the army at Fort Lewis, Washington, on October 15, 1971, are difficult to describe. On the one hand, no one was screaming or shooting at me, and I was completely free to do as I pleased. After three years of institutional control over nearly every aspect of my life, I needed no one's permission to go where I wanted to go and do what I wanted to do with whomever I wanted to do it. On the other, I was three thousand literal miles from home and a million figurative miles from my envisaged vocation as a field biologist. My rare opportunity to serve in the front lines of two wars—a cold one in Germany as a firing platoon commander for a "tactical" nuclear missile battery (black humor among us for postfiring procedures went like this: [1] bend over, [2] put your head between your legs, and [3] kiss your ass goodbye) and a hot one in Vietnam as an aerial observer for an artillery battalion in the central highlands—completely wasted.

The rank, combat efficiency reports, medals, commendations, and hard-won experiences added up to nothing more than a giant black hole in the middle of my embryonic professional résumé.

Yadda, yadda, yadda. Time to start over.

Two days later I was in the bosom of my family at my mother's place on the shores of Lake Chautauqua in western New York state. Certainly, it was great to be home, although my kids did not know quite what to make of me, especially my son, John Jr. (Jay), ten months of age, whom I had seen for only a week or so in his short life when I was on leave in July. I had no illusions regarding the task ahead. Snatched out of graduate school at the University of Minnesota by my draft board after a single quarter of work in the fall of 1968, my academic prospects were bleak.

My little family and I left New York for Minnesota in mid-December, arriving to take up residence in our new home at the fag end of the year. There was no snow as I remember when we took the Dale Street exit off I-94 on that gray afternoon, but the icy winds scuttling dead leaves along the grimy streets were as expected for the season in the land of ten thousand lakes.

My first task was to get myself reinstated as a student, a job that required a beneficence from the head of the Bell Museum graduate program.

I should explain a little about the Bell Museum, I guess. For centuries, the great natural history museums of the world, such as those of London, Paris, Vienna, and Berlin, led scientific inquiry. That tradition continued in the United States, beginning with the establishment of the Smithsonian Museum of Natural History in 1846. Nowadays we tend to think of museums as places to put dead stuff, but such have always been the leading scientific institutions, complete with faculty, students, and academic and research programs. Most if not all US states have natural history museums and associated academic programs at their premier universities.

The Bell Museum is a sterling exemplar of this tradition. Established in 1872 by the state legislature, it was home to the ecology and field biology research for the University of Minnesota. The building itself was a beautiful and imposing stone structure sitting on its own large lot on Church Street in the heart of the Saint Paul campus. During my time, it housed one of the top field research faculties in the country.

Attempting to return to this august institution after my hiatus thanks to the war was daunting. Colleagues who had been starting their graduate careers like me in 1968 were nearing completion—competing for, and landing, top jobs around the country at places like the Smithsonian, Rutgers, and the University of Florida. My major professor, Dwain Warner, who had been at the top of his game with a wealth of grant money and a stable of top-notch students, had fallen down a rabbit hole thanks to career disappointments and personal issues.

It was 10:00 on a bleak wintry morning when I arrived to discuss my reentry into the graduate program, the majestic oaks fronting the building standing as "bare ruined choirs where late the sweet birds sang," just as the bard says.

As I have noted, the museum was an impressive structure. You entered the high-domed gallery via a broad stone portal with massive doors. Stairs to the right led up to the inner sanctum—floors housing labs, offices, and specimen collections and centering on an impressive suite for the director and his immediate minions overlooking the entryway.

I arrived at his office promptly and informed the director's secretary of my appointment. She arose and went to the door of his office to announce me. On receiving an apparent positive response, she ushered me in to the Presence.

Harrison "Bud" Tordoff was the new head of the museum as well as chairman of the Department of Ecology and Behavioral Biology graduate program (housed in the museum). He had replaced the grandfatherly Walter Breckinridge during my absence. Breck, as we all called him, had considered the museum students as his children—wayward at times, perhaps, but always deserving of his loving attention. He treated us as nascent professional members of the field biology club, a status emphasized by his awarding of a master key to each new initiate that opened every door in the museum, including his own. During my personal calamity on being drafted out of his care back in 1968, he had allowed me to store all our household goods in the basement of the museum until my brother and I were able to drive out from New York a month later to get them.

Dr. Tordoff was a different species altogether. I had met him on my visit to potential graduate schools in early 1968, back when he was head of field biology at the University

Figure 1-1. The "fam" in 1972: Bonnie, Brigetta, Jay, and me.

3

of Michigan. He had turned down my application to his program then, and neither the change of venue nor the intervening years had altered his opinion of my abilities.

He gave me a frosty smile as I entered and invited me to take a seat. He apparently remembered me and got right to the point. While he would allow my reentry, it was only because he had to—it was against policy to deny a student return to a position held in good standing prior to being drafted. However, he continued, "You would have no chance of being accepted here today, Rappole. Your qualifications simply aren't competitive. You will be allowed to complete your master's, if you can, but certainly not to continue on for a PhD. Also, since your major professor is basically AWOL (Tordoff had been an army pilot in World War II and knew the lingo), you will have to find someone else to fill that role if you are even going to complete an MS."

With this welcome, my wife, Bonnie, and I, along with our kids, Brigetta, age three, and Jay, age one and a half, settled into our third-floor apartment in an old brownstone on Summit Avenue in downtown Saint Paul and began our new lives.

BACKSTORY

Whither thou goest, I will go; and where
thou lodgest, I will lodge: thy people shall
be my people, and thy God my God.

—Ruth 1:16

The reader may be forgiven for wondering how it was that we arrived at this difficult juncture. Robert Burns said it pretty clearly: "The best laid schemes o' mice an' men gang aft agley." The following brief summary may help one understand how a pretty reasonable basic plan went seriously *agley*.

Despite the fact that the majority of the smartest people in our high school class were women (fifteen out of the top twenty, plus the valedictorian), the work options for young females in 1964 were still limited—secretary, teacher, or nurse, for the most part. Bon chose nursing, attending Jamestown Community College on a New York State Regents Scholarship, graduating in 1966, and immediately passing her state boards as a registered nurse.

My parents decided that since Bon was now a professional who would have no trouble finding work, we could marry. However, there were a couple of bureaucratic hoops to be negotiated. For one thing, Colgate (all male at the time) required that students wishing to marry seek permission from the dean of students, which I did.

"Do you love her?" asked Dean Griffith.

"Yes, sir, I do," I responded.

"Well, I guess it will be all right, then." (Sheesh—talk about in loco parentis.)

Also, although women in New York state at that time reached majority at eighteen (Bonnie was twenty), men did not achieve that status until twenty-one. As a result, my mother had to sign for me to get our marriage license.

We married at St. Luke's Episcopal Church at 2:00 in the afternoon on June 11, 1966, after the end of my sophomore year. After a week's peripatetic honeymoon, Bonnie went to work at the local hospital in Jamestown while I dug ditches for the gas company and

Figure 2-1. Bon and me launching into married life in the getaway car, driven by my younger brother and best man, Robert. Hiding one's personal vehicle was a necessary precaution in those days because part of mid-twentieth-century nuptial rituals included the playful "decoration" of the happy couple's car by "friends."

drove a backhoe for a local concrete firm owned by a high school friend. We lived with my parents on the grounds of Chautauqua Institution for the summer and then headed out to begin our married life at Colgate in Hamilton, New York.

My father was a surgeon, my grandfather was a surgeon, and my older brother, Bert, was a surgeon (now retired). I had been accompanying my father to local hospitals and clinics to observe various operations from a tender age. So, naturally, I was premed at Colgate. Bon assumed, of course, that medical school would follow my undergraduate years, followed by internship, residency, and then settling down with a nice practice back in good old Jamestown, like my brother, where she could work as my nurse until there were too many kids, requiring her place at home.

These plans changed radically (disastrously, some might say) thanks in large part to Colgate's January studies period. The concept was that during the monthlong break between the fall and spring semesters, a special academic term would be set aside for intensive coursework in the discipline of one's own choosing. Faculty had on-campus offerings, but students were encouraged to develop their own courses to be administered by a faculty member of their choice. During January 1966 (prior to my marriage), my parents were taking a month off at a house they had rented on the beach in St. Croix, US Virgin Islands. I decided that I would devise a course of my own to be done on the beach in St. Croix rather than in bleak Hamilton, where the skies alternated between various shades of gray throughout the entire winter.

I proposed to my excellent advisor, Dr. Robert Goodwin, that I should do a field study of the birds of St. Croix. He agreed, and I began to prepare for my trip by reading everything I could find about what had previously been done on the topic. Based on this review, I knew that a major contributor to that knowledge was George Seaman and that he worked at the extension offices of the US Department of Agriculture (USDA) on the island.

When I arrived, my parents provided funds for the rental of a jeep, and I was off. My first visit was to Mr. Seaman—a delightful and exceptionally knowledgeable man who was happy to share his vast familiarity with the avifauna and best places to visit. I spent the next month more or less continually in the field, checking in regularly with Mr. Seaman to discuss my observations.

Back at Colgate, I wrote up my findings for Dr. Goodwin. He gave me an "Honors" evaluation (there were only three grades—pass, fail, or honors) and suggested I submit the material for publication in a professional journal. I did, but it was rejected for lack of specimen documentation for the new records.

Nevertheless, I was hooked. I wanted to be a field biologist. But who pays you for that? The answer to that rather basic question took a number of years to discover.

Now married in my junior year in the fall of 1966, I decided that if I was going to be a physician, I might as well get after it. Few applied for medical school a year ahead of their senior year, but I thought it could be a kind of test. If I were really meant to be a doctor, I would be accepted, and if not, I wouldn't be, and I would then be free to follow other avenues—weak reasoning, I have to admit. Anyhow, after fall semester, I applied to the University of Buffalo (UB) Medical School, where all of my relatives had matriculated.

To my great surprise, I was granted an interview, which was more or less pro forma for admittance at that time. (I should not have been surprised; despite my low B average and C in organic chemistry, I had done well on my medical college admission test [MCAT] and was a triple legacy at UB.)

In any event, I went to UB for the interview, which did not go that well. The first question the professor asked was "Why do you want to be a doctor?" I mean, who knew he would ask that? Talk about left field! I answered pretty lamely that my father was a doctor, my grandfather was a doctor, my uncle was a doctor (although a University of Rochester graduate), and my brother was a doctor. What else could I do? My interviewer was patient, but he gamely explained that he felt my reasons at present were insufficient. I should go back to Colgate, get a better grade in my second semester of organic, think more about my calling, and reapply next year.

Very good advice. And quite helpful. I took it as a sign—perhaps not from heaven but from somewhere important. Returning to Colgate, I had a long discussion with Dr. Goodwin. "Is it possible for a married man with a family to earn a decent living as a field biologist?" I asked him.

"Of course!" he said. "What do you think I am? My doctorate is from Cornell in vertebrate zoology, but my specialty is mammals, and my thesis was on bats. Lots of college professors are field biologists by training."

"How do I prepare for such a venture?" I asked.

"Do the best you can in the courses in which you are already enrolled. Get into Dr. (Roger) Hoffman's experimental biology courses if you can. They will be good training for how to go about identifying, testing, and answering important biological questions. Apply for summer positions with field biologists looking for help. Prepare well for your graduate record exam (GRE), and most importantly, consider what kind of thesis you would like to do, find the professor best qualified to guide you, and arrange to visit him."

With that advice in mind, I dropped medical school from my thinking. Bonnie was not upset about this. She knew that Dr. Goodwin was right. College professors led normal, comfortable lives. One thing that neither of us paid much attention to at this point was the war. True, in the spring of 1967, it was heating up fast, but I had my student deferment, which would take me to the end of my time at Colgate. And I had taken a national test during my sophomore year, my score on which deferred me through graduate school. Also, I was married—another deferment. What we (and few others) understood was that deferments are ephemeral. Except for the one for medical students. They were deferred through their entire training (which got my older brother through his five-year surgical residency at Bellview Hospital in Manhattan), and that one (that is, deferment for medical training) lasted through the entire war.

Dwain Warner, my thesis advisor at the University of Minnesota, was a remarkable individual with a first-rate intellect camouflaged by a Falstaffian zest for life and love of a tale well told. Bonnie and I had first met him in January 1968 on our "Donner Party" tour visiting possible graduate schools. Starting out from Jamestown, we had headed west to the University of Michigan—where, as I have mentioned, Dr. Tordoff told me I did not have the "right stuff"—then on to the University of Wisconsin and the University of Minnesota, experiencing multiple blizzards and vehicle breakdowns along the way.

We passed the night before our scheduled visit with Dwain at a motel located just east of the Wisconsin-Minnesota line. The temperature dropped to twenty-seven below, and when I went out to try to start the car the next morning, I found slush in the radiator. Bonnie had had about enough by this point and told me, "I've learned a lot about you on this trip." (I may have been a little short in response to her question about the car's status.) Nevertheless, we trekked on across the frozen prairie.

Only forty miles remained between our motel in Baldwin, Wisconsin, and Dwain's office at the Bell Museum on the Saint Paul campus, but with the car heater not working, it seemed a lot farther than that. In fact, we had to stop at a 1930s-era tire repair shop along Highway 12 to beg the proprietor for a few moments by his woodstove until we defrosted enough to finish the trip. (He seemed a bit perplexed and amused. "These Easterners," he must have thought, "ain't likely to survive up here when it gets really cold.")

We met Dwain at noon, a couple of hours late for our appointment, but it didn't seem to bother him much. Unlike my earlier encounters with professors from other universities on the trip, he greeted us enthusiastically and said that we should plan on joining his program, probably to do work involving the radio-tracking of a bird or mammal species of our collective choice. Dwain was a pioneer of the method, working with an engineer, Bill Cochran, to develop the technology and establish the university's Cedar Creek tracking center, complete with towers and monitoring machines—the first such center of its kind in the country for the radio-tracking of wildlife species.

As a result of this interview, we left the Twin Cities on our southern swing of visits (Purdue, Indiana State) feeling pretty good—although we never turned off the car for the next couple of days until the temperature finally got above freezing in southern Indiana. So after graduation from Colgate in May 1968, I started life as a graduate student and curatorial assistant in the bird collection at the University of Minnesota's Bell Museum.

When we arrived that summer, Dwain, a full professor, fifty years old then, was at the height of his powers. In a department that boasted five of the top ornithologists in the country—Walter Breckinridge (who was about to retire as director of the museum), Dave Parmelee, Kendall Corbin, Frank McKinney, and himself—Warner stood out as the one most capable of getting grant funds and attracting the best graduate and post-doctoral students, including such luminaries as Bob Dickerman, Byron Harrel, Gene

Figure 2-2. Professor Dwain Warner and friend. Photo courtesy of Dwain Warner.

Lefevre, and Lew Oring. Unlike most of his fellow academics, Dwain found it easy to talk with the Twin Cities money people—folks like the Bells, Pillsburys, and Daytons. "They don't make pockets in shrouds," he'd chortle, and they evidently got the message because they provided several little (to them!) pots of money to support his projects and students. He knew the science, and he knew how to make it clear and compelling to nonscientists. He had intelligence, imagination, and an ability to communicate complex information as part of a good story.

Unfortunately for us, our Minnesota idyll came to a sudden end after the completion of the 1968 fall quarter, when I was drafted into the US Army. Bonnie headed back to Jamestown with our daughter of one month, Brigetta, and I headed off to Fort Dix, New Jersey, to learn how to kill.

The next three years included six months in New Jersey, ten months in Oklahoma, eight months in Germany, and the last year or so in Chautauqua at my parents' place for Bonnie and Brigetta (and our son John Jr. for the last seven months) while I was in Vietnam.

CHAPTER 3

INITIATION

Wherein I Begin My Life as a Field Biologist

The crop of graduate students I joined was indeed impressive, including people like Doug Mock, Joanna Berger, Henry Kermott, George Barrowclough, Scott Derrickson, and Joe Wunderle, who would go on to have brilliant careers in research and academia. They scoffed at the droves of GI Bill vets who were, as one of my colleagues observed, "swarming the campus in their ratty field jackets, scuffed combat boots, and bits of fatigue uniform, grasping shiny briefcases whose sole contents is an old *Playboy*."

I felt ancient and knew that contempt for returning veterans was largely well earned. We were older. We had seen a lot, most of which was not helpful preparation for a life in academia. We had different views and values, we weren't intellectuals, and some of us even had children. Sex was all well and good, but families were considered an affront to reason. "Haven't you read Malthus and *The Population Bomb*? Too many kids is the main cause of the world's problems!" Only one class of individuals was held in lower esteem by my fellow grad students than the family man—namely, altruists, which, of course, is what warriors are. They simply smirked disdainfully at my ignorance when I pointed out that people like themselves, who chose not to have progeny, were being equally selfless (or more so) from an evolutionary perspective.

That first quarter back in the spring of 1972 was tough. The classes—systems ecology, biochemistry, and statistics—were OK, but when Bonnie and I paid a social call on Warner and his wife, Gloria, I knew I was in just as much trouble as Tordoff had suggested. Dwain was pretty much incoherent throughout the visit. Gloria tried to be reassuring, alluding to his upcoming enrollment in the Hazelden Clinic, but the signs were bad. I had to accept that unless he could get back to being a functional professor, my prospects at Minnesota were grim.

A graduate student without an advisor is like an orphan chick. Survival is unlikely. I had no relationship with any faculty other than Warner on my return in 1972. My best hope seemed to be to do well enough in my systems ecology class to impress the professor sufficiently with my potential for him to accept me as one of his master's students. Accordingly, I worked hard to be at the top of the class, and toward the end of the quarter, I went to see if I could get him to take me on. He was sympathetic but turned me down, explaining that he had too many students already, and my area of interest wasn't really systems ecology. I was devastated and figured that our days in the Twin Cities were numbered.

Then a miracle happened. Warner returned from Hazelden sober and started coming into his office at the museum after a year or so of unexcused absence, reassuming his professorial and curatorial duties (he was curator of birds at the Bell Museum as well as a full professor in both the Department of Ecology and Behavioral Biology and the Department of Zoology). Of course, his entire group of top-level graduate students and postdocs was gone along with all of his funding, but at least he was back.

With all of that behind us now at the end of the spring quarter of 1972 and Dwain seemingly back in the saddle, it was with considerable relief that I started life as a real graduate student, working diligently with him to try to come up with ideas for the three "minitheses" I would need to complete a coursework master's by the end of the year.

One thought we had was to do statistical analyses of the TV tower kill data set compiled by Vince Haigh. Vince had been a doctoral student of Dwain's a few years back. His doctoral thesis was to have been on the population ecology of passage songbird migrants as revealed by individuals killed during nocturnal migration by a two-thousand-foot TV tower located in Eau Claire, Wisconsin. Accordingly, he had made many visits to this tower to collect hundreds of birds that had collided with its guy wires, recording the date, species, sex, and age for each one found. A final analysis had never been accomplished because Vince had been offered a faculty position at the University of Wisconsin, Stevens Point, before completing his degree. With his main goal of landing a job achieved, he never finished.

I should say a bit about TV towers here. Most people, if they have thought about them at all, likely consider them an atavistic eyesore or, in the context of birds, as one more threat on their hazardous travels. For the field biologist, however, they represent a sampling technique providing a sensational snapshot of the composition of a night's migration activity.

I had never thought much about bird migration, but when this opportunity turned up, I jumped at it. In 1972, there were no personal computers, so if you wanted to do a computer-assisted statistical analysis, you had to enter the data on punch cards and then take them down to the university's computer center to have them run, which is what I did with the tower-kill information.

The results were fascinating. Sample sizes were large enough to get a decent idea of what was going on for about ten species, including birds like the Red-eyed Vireo, Swainson's Thrush, and American Redstart. The data showed that the timing of migration differed significantly for adult males, adult females, and young birds hatched during the immediately preceding breeding season for all of the species in the analysis.

The Swainson's Thrush data were particularly intriguing. This bird has several subspecies—that is, populations that breed in different parts of the species' range that can be identified by plumage. It appeared from our data as though at least two of these subspecies migrated through Wisconsin and that passage timing differed among the age and sex groups of the different subspecies. We did not have enough specimens to perform a statistical analysis of the question. Nevertheless, the information we had implied that there was a great deal more to be learned about the topic of migration from passage transients sampled in this way.

Exposure to the TV tower data had gotten me excited about migrant population and behavioral ecology as a possible PhD thesis topic. But field studies are and were expensive, and few organizations provided funding for this sort of research. Dwain advised putting together a tangential project, one that would attract funders with the promise of information they could use for their purposes that would allow us to pursue the questions of interest to us at the same time. This approach was a skill of Warner's, one that is absolutely essential for those wishing to develop a successful career in field biology, the basic elements of which excite few investors. In this instance, Dwain suggested that I write a proposal to study the migration patterns of the Red-winged Blackbird and submit it to the USDA. Redwings cause significant damage to sunflower and other seed crops in the upper Midwest. My job would be to make the case to the ag people that learning about migration would help them understand how to control redwing populations, while I would use the same data to answer basic questions of migrant ecology.

Truthfully, I was having some difficulty motivating myself to put together this kind of pitch when a serendipitous event occurred. A South African professor visited the museum to give a seminar on his research. After his presentation, Dwain and Gloria hosted a wine-and-cheese (in reality, tonic and bull balls, Dwain being on the wagon) event at their home where faculty and graduate students could meet and talk with the visiting nabob in an informal gathering. Ideas were the most common topics of conversation at such events, but money was not far behind. Nearly all graduate students are, by definition, starving, both literally and figuratively. My colleague and friend Joe Wunderle, for instance, spent his early graduate career at Minnesota bunking on the floor of the botany lab and subsisting on baloney sandwiches. New students in particular devote immense amounts of time desperately searching for money to support themselves and their research. Discussions with the South African professor gravitated naturally in this direction, and it turned out that he had some interesting information. As part of his US lecture tour, he had stopped at Texas A&M University and been invited to visit the campus of the Welder Wildlife Foundation, a private institution located on the Gulf Coast about forty miles north of Corpus Christi that provided support for a number of A&M wildlife students. He said that the place was actively recruiting proposals. They had funds and wanted to share!

Dwain already knew about Welder. He had stopped there in 1954 on his return trip from his sabbatical in Mexico when it was first being established and met its legendary director Dr. Clarence Cottam at that time and again at the formal dedication in 1961. It had not occurred to him that Welder might be a funding source for graduate students from schools other than A&M, but the South African said they were considering proposals from anywhere. So I got down to work, as did others in our department.

We decided to base our proposal on questions arising from the TV tower data. Dwain suggested that it would be wise if we sited a portion of the research at Welder, where work on transient birds could be located, supplemented perhaps by the investigation of migrants wintering in the tropics of eastern Mexico. The actual writing was my job, of course, so I immersed myself completely in the rich literature on the ecology of transient and wintering migrants, topics about which I knew precious little prior to this undertaking.

Dwain was always optimistic about such things, but personally, I was stunned when Dr. Cottam contacted us in November 1972 with the news that our proposal had been funded. There were a couple of provisos: First, the two technicians included in our budget as my assistants would have to be graduate students, funded by Welder, who would work on their own projects in addition to helping me. Second, one of these students would have to be a Mexican, since a major portion of the research would take place in that country.

We readily agreed to the suggested alterations in our research plan. Warner's technician, Dick Oehlenschlager, a native Minnesotan from a farm near Nimrod who had nearly earned his bachelor's degree from the university a few years previously, agreed to become a graduate student for the purposes of the proposed work. Dick was not really anyone's idea of an academic—least of all, his own. His favorite pastime when he wasn't at the prep lab was to sit at home in front of the TV with a tray in his lap skinning birds for the museum collection. However, he was perfect from Warner's perspective. Dwain's ambition for the project was to build the museum's collection into one of the top repositories of information on Mexican birds in the world, a job that would require a large amount of collecting (killing) and skinning, which made Dick, who was an excellent shot and highly skilled (and speedy) preparator, invaluable.

As for the Mexican student, Dwain contacted his old friend and colleague Allan Phillips, who was then a professor and curator of birds at the Instituto de Biología of the Universidad Nacional Autónoma de México (UNAM) in Mexico City, and asked his advice regarding the best young Mexican student in ornithology. Without hesitation, Allan recommended Mario Alberto Ramos Olmos, thereby initiating the professional career of Mexico's greatest ornithologist and international leader in conservation biology.

Naturally, I attributed the success of our Welder proposal to the brilliance of my oeuvre, while Dwain assumed that it was because Dr. Cottam remembered him fondly from their previous brief encounters. It was sometime later that we found out the most probable reason for our triumph. In early 1972, tax laws for nonprofit organizations like Welder had been changed, requiring them to spend 85 percent annually of the mineral royalties and investment earnings on the stated purpose of the organization, which, for Welder, was grants to graduate students. This change was retroactive, and since for some years Dr. Cottam had been putting more of the earnings into the endowment than was allowed by the new provision, Welder had a slug of money they had to disburse as grants. The timing of our submission likely had more to do with our selection than any other factor.

THE GREAT ADVENTURE BEGINS

Come, my friends,
'T is not too late to seek a newer world.

—ALFRED LORD TENNYSON, *ULYSSES*

The project began in January 1973. Dwain and I decided to make an exploratory trip down to Welder late that month and then to travel into Mexico to scout for possible study sites and meet and talk with leading Mexican field biologists to get their advice on where and how the work might be done (and give Warner a delightful trip down memory lane).

My wife, Bonnie, and our two children—Brigetta, now four years old, and Jay, two—accompanied us as far as Welder, where they were to remain for a month and a

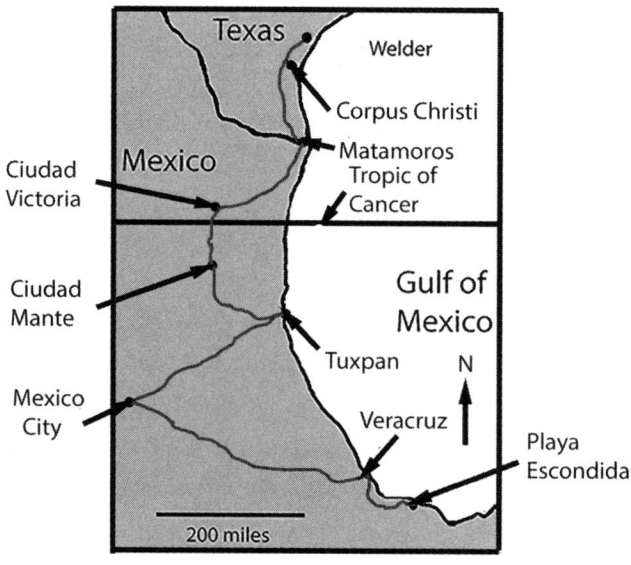

Figure 4-1. Route followed on the exploratory trip to Mexico, February 1973.

half while Dwain and I, along with Oehlenschlager and Paul Haemig (an undergraduate at Minnesota then, now a professor at Linnaeus University in Kalmar, Sweden) continued down into Mexico.

Prior to our departure, Dwain advised me that I should not consider bringing my family on this trip or on any other part of my fieldwork. His first wife (who had passed away a few years previously) never accompanied him in the field. She stayed at home in Minnesota and raised their six kids. I explained to him that while I appreciated his perspective, after three years of having the US Army tell me where I had to go and with whom, I would take my family wherever I wanted.

Leaving the Twin Cities, we set out for Texas in separate vehicles, with Dwain and his crew in a university carryall (U of M gopher colors of maroon and gold) and my family in our Chevy II station wagon. We met up at Welder in late January and spent a few days getting to know the Welder staff, familiarizing ourselves with possible research sites on the refuge, and for me, getting Bonnie and the kids comfortably established in the student dorm.

Figure 4-2. The kids, Brigetta and Jay, settling in at Welder, January 1973.

We set out from Welder for Mexico on February 4. It was a clear, bright morning when we headed south on Highway 77 across the empty grasslands of the King and Kenedy Ranches, fields of cotton and sorghum of the Rio Grande delta, and finally into Mexico at the border town of Matamoros at the northern edge of the state of Tamaulipas.

The Mexican excursion was a miraculous experience for me. Interestingly, although many scenes are still vivid in my mind, I can tell from rereading my field notes and accounts of the trip by Paul, myself, and Dwain (chapters 10, 16, and 19, respectively, in Kevin Winker's book *Moments of Discovery*) that memories of events are not necessarily the same for all participants. Nevertheless, I guess we agreed on the most important parts, which focused on locating appropriate study sites.

Mexico is not the United States. It isn't now, and it certainly wasn't then. I have crossed many international borders in my long career, but none has presented so striking a contrast between worlds in such a short space as I experienced on that trip, exiting the "modern" world of Texas—with its broad, beautifully paved highways, soaring cities, and neat towns with fully stocked stores and trim, well-watered, well-lighted, and air-conditioned residential areas—and entering into a time and culture much the same (to all appearances) as during the revolution a half century ago, where the roads were mostly dirt and the commonest mode of travel was on foot.

The main route south out of Matamoros for all travelers in 1973 was the famous Pan-American highway, an unprepossessing stretch of pocked asphalt extending across all of Mexico and Central America before petering out in the Darien wilderness of southern Panama. The scenery during the first Mexican portion of the trip was much like the Texas Rio Grande Plain, with vast vistas of sorghum fields spreading out to the horizon in every direction. But after a hundred miles or so, you leave agriculture behind and enter ranch country—no settlements, no people, and thorn forest as far as the eye can see.

In Texas, ranchlands are and were maintained as grasslands by the government-subsidized clearing of woody plants or "brush," with herds of beef cattle specially bred for the brutal South Texas climate, like the beautiful roan Santa Gertrudis (three-eighths Brahma, five-eighths shorthorn), a common sight. Not so in 1970s Tamaulipas, where the cattle were pretty much left to fend for themselves in the inconceivably dense stands of mesquite, prickly pear, yucca, blackbrush, and so forth, the principal sign of their presence being dusty paths through the undergrowth dotted with occasional piles of dried dung. The actual animals, scruffy beasts of indeterminate parentage, were seldom seen.

This was an unpromising scene to a Texas rancher's eye, perhaps, but thrilling to me at the time and still today. For the naturalist, the Tamaulipan thorn forest is a

Figure 4-3. Tamaulipan thorn forest north of Ciudad Victora.

fantastically rich environment. This much was obvious to me even as inexperienced as I was and flying along at seventy miles an hour in the carryall. It was early spring. The yucca was in bloom, and the mesquite was leafing out in a delicate green. We stopped once to pick up a dead javelina for a museum skin, and the cacophony of strange birdsong and the waft of fragrances from the chaparral were overpowering.

The proximity to his old haunts and the need to locate a place to spend the night stimulated Dwain to a recounting of one of the more memorable stories from his considerable repertoire, "A Shot in the Dark." Everyone who has ever spent time in the field with Dwain has heard this anecdote on multiple occasions. In fact, Kevin Winker included a version of it in his aforementioned book of field biologist narratives. This telling was the first time for me that I can recall.

I will begin the saga by providing some context, which would have been clearly understood by fellow field biologists but perhaps not so much by the average reader. At the time, probably the late 1950s (Warner was never completely clear on this point), Dwain was doing fieldwork in the highlands outside Cuernavaca. After spending a week in the bush, he decided he needed a meal fixed by professionals and a room for the night to catch up on his specimen preparation and field notes. To explain a bit, the daily routine in the field for vertebrate zoologists inventorying an area begins at dawn, when they head out to check traps set to catch small and medium-sized mammals the night before, armed with a gun for shooting birds, mammals, reptiles, or amphibians ("herps") encountered by chance. After running the check, they then return to camp to catalog, label, and prepare the dead animals for preservation as museum specimens—usually as skins stuffed with cotton for mammals and birds or simply jarred in formaldehyde for herps. Once this processing is complete, the notes for this segment of the day are written, including the date, time, weather, habitats sampled, species seen but not collected, and any interesting behaviors observed or anything else of possible use or interest that might have occurred. These procedures are repeated throughout the day and often late into the evening—depending, of course, on the number of specimens collected. After several days of this routine, most of us begin to fall behind on the summary portion of the field notes, the last things written before sleep where you try to put together what you think you might have learned from the day's activities. These summaries are not too tough if you can sit down at a table in a lighted room to prepare them. Personally, I find it more difficult sitting on a rock, hunched over the fire, trying to hold the flashlight and write at the same time. Anyhow, Dwain took a break and headed for the nearest hamlet with a hotel.

The one chosen, I assume, was standard for the period: a rectangular concrete structure surrounding a packed-earth courtyard, which served as a guarded parking area for vehicles and pack animals. Rooms likely were entered via a doorway off a long hall, each about twelve by twelve with unadorned concrete floors, walls, and ten-foot ceilings, from the center of which a naked bulb dangled to about seven feet or so at the end of an electrical cord. Probably a single window opened onto the courtyard. Furniture likely included a low, shaky bed with a shabby mattress and pillow, a deal table with a straight-backed wood chair, and a wooden commode in the corner with a porcelain bowl and pitcher of water on top and a cabinet containing a chamber pot underneath.

So now you have the situation and setting. Let the story (as Dwain would tell it) begin:

"I got to the room after supper and went straight to work preparing specimens. It had been a good collecting day, so I had quite a few to do, finishing up at about ten in the evening. I then went to work on my notes and finished those by eleven or so, when I threw my poncho over my mattress and pillow (never make it easy for the bedbugs and other vermin to get you), set my pistol down on one side of the pillow and my flashlight on the other, lay down, and went directly to sleep. I was awakened sometime later by a scruffling sound coming from over by the door. Fearing possible intruders, I grabbed and switched on the flashlight with my left hand while retrieving the pistol with my right. The light revealed a startled wood rat with pygmy-owl remains dangling from its mouth. I shot, of course, just as I would have at camp. Jesus, what a noise! A pistol going off in a concrete room sounds like a mortar round landing in your foxhole, even when loaded with bird shot. I shut off the light immediately and waited breathlessly for all hell to break loose. Nothing. Not a sound—then or for the rest of the night. *Must be a pretty regular occurrence*, I thought, *not worth investigating or risking your neck.* The next morning, I packed specimens and gear in my duffel bag along with my holster and pistol, went down, and checked out. No questions, no problems. There was not much left to salvage from the pygmy-owl, but the rat was OK. I gave it to Bernardo Villa, curator of mammals back at the Instituto, who was delighted to have a Morelos specimen of the southern highland subspecies, *Neotoma mexicana torquata*."

As a barely relevant aside, I will contribute a similar story. In the mid-1970s as part of our fieldwork on the Welder project, our Mexican student representative, Mario Ramos, and I were seated in an open-air restaurant patio with a thatched roof in the lakeside town of Catemaco enjoying a cool beverage when an *Empidonax* flycatcher flew in and landed in the rafters. There are only two common species of this group of flycatchers that winter in southern Veracruz, and we both recognized this bird as belonging to neither. Mario's thesis work focused on the identification of wintering areas of migrant populations, and it seemed likely that this bird, a probable member of the confusing Willow-Alder species complex, might be able to contribute some interesting information. Since we were the only customers, Mario stepped to the bar and asked if it would be OK if he shot the bird. The surprised bartender said that he supposed so, whereupon Mario dropped the bird neatly on the floor beside us. It turned out to be a Willow Flycatcher—an unusual and enlightening specimen—identified for us by Allan Phillips, world expert on the group.

Warner always loved that story.

Anyhow, back to our 1973 exploratory trip.

It was getting dark when we pulled up in front of a likely *posada* on the *zócalo* (town square) in Ciudad Mante. Warner stopped the carryall and said, "OK, John. Get us a couple of rooms."

Jeepers! I took French in school and had about half a day's worth of Spanish picked up on the trip, but I saw the point. If I was going to do a PhD project on my own in Mexico, I had better learn how to communicate. So I said, "All right. What's the word for *rooms*?"

"*Cuartos*," he said. "Just go in and say '¿Hay cuartos?'"

Which I did. The clerk said, "Si, hay. ¿Cuántos?" And I said, "Dos." He asked for my visa, passed me the register, and I signed in. No problem.

Figure 4-4. Warner with a bobcat near Cuernavaca in 1953. Photo courtesy of D. W. Warner.

We parked in the hotel garage (only an insane person would park in an unenclosed, unguarded place) and headed for our rooms; Dwain and me in one, Dick and Paul (now dubbed Pablo) in the other. After getting ourselves installed, we headed out to find something to eat.

In my opinion, Mexico is the best country in the world for diners. There's always someplace, no matter how small the town, and it's always good and usually great. And even if there are only a few huts, just stop at one and ask. Most folks are happy to cook you up some beans, rice, and chili with tortillas in short order for a modest fee. Beef, chicken, or kid might take a little longer.

After a delicious supper (*cena*), Dwain was still wound up and in storytelling mode on his first night in Mexico since the early 1960s, so we set out on a stroll to explore the offerings on the *zócalo*.

Like most others that I have visited, this one was lined with shade trees, mostly live oaks in this case, filled with hundreds of Great-tailed Grackles, which fly into the town from the surrounding fields in the evening. My dictionary defines *roost* as "a place where birds go to sleep." But this seems wrong, at least for these birds, who appeared to spend most of the night pooping and shrieking.

In any event, they provided powerful sensorial stimuli, prompting Dwain to reminisce about *zócalos* of old from his visits of the 1940s and '50s. Back then, he said, it was traditional for young ladies of marriageable age to walk clockwise around the square accompanied by one or more *dueñas* while swains interested in courtship similarly would walk, but in counterclockwise fashion. After several passes of mutually agreeable inspection, the young man might invite the young lady to sit with him for a moment on one of the park benches (carefully cleaned of grackle droppings, one supposes) to become better acquainted. An apocryphal tale, no doubt, like many of Dwain's, but entertaining and properly set to the moment.

We did not observe any obvious activity of that sort during our stroll, but we did find a marvelous *taberna*—an old saloon with battered batwing doors opening into a large room floored and paneled in dark wood. Overhead was a large, lazy fan, and against the far wall was a chest-height bar fronted by high stools and a footrest brass rail. Behind the bar was a long mirror and shelves holding a variety of *bebidas*. A three-man mariachi band was playing energetically near the right-hand side of the bar. Just the kind of place you might expect Cisco and Pancho to belly up for a couple of *mezcales* neat after a hard day chasing the *patrón's* hired thugs. There were only a few customers, sitting separately and nursing their drinks at bare tables as we entered. No women.

Despite our obvious gringoness, we were welcomed warmly enough by the *camarero*: "Buenos tardes, señores. ¿Qué quieren a tomar?" Warner ordered a Coke, as did Pablo. Dick and I ordered a draft.

Although I had traveled a good bit with my family in Europe and as a guest of the US Army, I had never experienced such a quick and total cultural immersion. Best of all was the band, which consisted of a guitar, *guitarrón* (bass guitar), and trumpet. The guitarist and bass player were unexceptional *campesinos*, but the trumpeter was special. He was dark featured and of medium height, shabbily dressed, hatless with shoulder-length black hair shot with gray pulled back in a queue. His left eye was covered with a black patch, from which livid scars extended toward nose and ear. His left arm was missing below the elbow, so he played his instrument with his right hand alone. His appearance seemed to embody for me the whole sad and brutal history of his people.

We sat for an hour rapt by the music, somehow jaunty and melancholy, joyful and sorrowful at the same time. We tipped the men well when we left, thinking there could be no more wondrous introduction to the spectacular country of Mexico.

Figure 4-5. Swainson's Thrush (above) captured in mist net (below).

I awoke the following morning to the mournful call of an Inca Dove, "No hope, no hope, no hope . . ." and scented air from the yucca in the hotel courtyard outside my window. Welcome to the tropics!

Ciudad Mante is located forty-five miles south of the Tropic of Cancer and ten miles or so east of the outer fringes of the Sierra Madre Oriental. As such, it served as a good takeoff point for our reconnoitering for study sites. The plan put forward in our proposal to Welder was to find pieces of wet tropical forest representing lowland and highland elevations. At each of these sites, we would set up mist nets to capture the migratory birds using these forests during the wintering period.

It would be appropriate here, I guess, to say something about the mist net. Famed ornithologist and longtime editor of the journal *The Auk*, Oliver L. Austin spent time in Japan following his service in World War II and learned there of the mist net and its use for the capture of birds (to eat). Returning to the States, he brought some with him, thereby introducing the ornithological community to this splendid tool. Since the time of that introduction, the mist net has become the most important apparatus for field studies after binoculars and the shotgun. The nets I used throughout my decades of migrant work were forty feet long and eight feet high and made of fine black nylon thread. The net is separated along its length by strings of thicker thread into four panels, or shelves. Excess netting is provided so that a loose bag hangs along the lower thread of each shelf. When the bird flies into the net, it tumbles into this bag and becomes entangled. Set properly, the nets are virtually invisible to birds flitting through the undergrowth.

Logistic reality dictated that our study sites be located near an all-weather road so that we could move with relative ease between sites separated by many miles. In Dwain's memory the roads proceeding from the coast at places like Tampico, Tuxpan, and Vera-cruz city up into and through the passes in the Sierra and across the Mexican Plateau to Mexico City would provide many possibilities for such sites. After all, he knew this part of the country well, having first visited in 1941 as a young PhD student with Cornell professors George Miksch Sutton and Olin Sewall Pettingill and fellow student Bob Lee (see Sutton's account in *At a Bend in a Mexican River*). Of course, that was the Mexico of John Huston's great film *Treasure of the Sierra Madre*, especially its opening scenes filmed in Tampico, a town located on the border between Tamaulipas and Veracruz. And Dwain had had many more years of work in Mexico since then. He had done a sabbatical in Mexico City in the early 1950s and been on several exploratory trips as an advisor to several students, including such luminaries in the field as Byron Harrell and Bob

Dickerman. It was based on these experiences that we had suggested in our proposal a transect of study sites at various elevations in the Veracruz portions of the Sierra Madre Oriental for studying wintering migrant biology.

As we set out that morning from Ciudad Mante, our plan was to proceed down to the coast at Tuxpan and then head inland from there to Poza Rica up into the mountains along the main "highway" through Huauchinango and Tulancingo and then across the plateau to Mexico City.

Sadly, dramatic changes had occurred in the years since Dwain had been in the region, and the forests that lined this route in his memory were ghosts—shades replaced by pasture with a few scattered trees in the lowland and midelevations and by coffee plantations and subsistence agriculture at the higher elevations, where, as Warner commented, "A guy could die falling out of his own cornfield." There were few sites within easy travel distance from the main road where even patches of forest remained, and these were usually along incredibly steep slopes bordering cataracts. We stopped to explore a few such places and found them dangerous in the extreme. I couldn't imagine setting up and running mist nets at even one, let alone several.

Figure 4-6. Paul Haemig (Pablo), Dwain Warner, and Dick Oehlenschlager lunching by the road on the plateau east of Mexico City.

With this information in hand, we went on to Mexico City to get our collecting permits and to meet with the top Mexican field biologists to get their ideas on where we might find the forests needed for our study.

It has been at least thirty years since I last attempted to drive a vehicle through downtown Mexico City, so perhaps things have changed. This first experience was similar to all subsequent efforts for me in terms of the sense of entering into complete chaos—a maelstrom of whizzing cars, trucks, and buses completely oblivious to any traffic control measures (stoplights, speed limits, etc.). Nevertheless, Warner navigated the carryall through the bedlam, with help from a Sanborn's map (the standard guide at the time issued along with your Mexican car insurance), onto the grounds of the elegant Hotel Escargot. Located just a block or two off from one of the main cross-city arteries, the Escargot was a lovely caravansary of old-world charm: comfortable rooms surrounding a courtyard with a fountain and gardens filled with flowers. There was even a concierge to help with such important activities as changing money, getting a cab, or calling various government offices and officials. Amazing, but typical Warner. He always knew exactly how to get stuff done the right way.

The next day, we set out to meet with the various officials and old friends and colleagues to see what we could find out about locating study sites and getting permission to use them for our project.

Figure 4-7. The "classical taxonomist" at work (Parkes 1963).

27

Our first such meeting was with Allan Phillips, who was working at UNAM's Instituto de Biología as a professor and curator of birds. Like Warner, Allan was Cornell trained, and although he was a few years younger than Dwain, they had overlapped there because of the delay in Dwain's career caused by his Second World War service (in the South Pacific). Subsequently, they had worked on projects together in Mexico in the 1950s and '60s. Allan was among the last and arguably one of the greatest of the classical avian taxonomists. His knowledge and contributions to understanding the distribution and variation in the avifauna of North America were unparalleled. Unfortunately for his academic reputation and career, the field of evolutionary biology was undergoing a radical transition in the '70s away from a focus on descriptive aspects involving the naming and classification of species using study skins, skeletons, and alcohol specimens to a broader attempt to understand all aspects of the evolutionary process, including not only taxonomic relationships and morphology but the interplay between organism and environment through the process of natural selection. As a result, and despite his unrivaled knowledge of new-world avian systematics, his work was largely ignored by leading evolutionary biologists of the day who, although happy to use his taxonomic expertise as a baseline, were openly contemptuous in print of his contributions. (It should perhaps be noted that he gave as good as he got, or better.)

This view of Allan as a professional curmudgeon was not improved by his acerbic personality. As Dwain said, "Allan is always positive—and sometimes right." Nevertheless, for those who understood the depth of his knowledge and expertise, he was an incredible resource. In addition, he was a devoted friend to Dwain (and many other outstanding field ornithologists, like Bob Dickerman, Mario Ramos, and Kevin Winker), ever willing to provide advice and assistance. He was also a brilliant raconteur and font of entertaining witticisms, recollections, and commentary. Like Rumpole, he could recite reams of poetry from memory as well as bawdy renditions of famous ballads such as *King Arthur and His Knights of the Round Table* ("Balls," said the Queen. "If I had two, I'd be King!") and *Dangerous Dan McGrew* ("Twists and shunts, and double bunts . . ." etc.).

We spent considerable time with Allan discussing our project, but his experience came mainly from work in the southwestern United States and western Mexico. He had not spent much time in eastern Veracruz. Nevertheless, he promised to help with species and subspecies identification, a critically important aspect for us, as it turned out, and one that no other ornithologist could have provided.

Our second contact was with another old friend from Dwain's Mexico City days, Bernardo Villa. Dr. Villa was Mexico's leading mammalogist and, like Phillips, a professor and curator at the Instituto. (He claimed Montezuma and Cuauhtémoc among his illustrious progenitors.) Trained at Kansas under E. Raymond Hall, doyen of twentieth-century mammalogy, his main area of interest was bats. His *Murciélagos de México* was the basic reference for the group in the region for many years. Unfortunately, he could not think of potential study sites in eastern Mexico that would match our needs but suggested we talk with famous plant ecologist Arturo Gomez-Pompa, then director of the Instituto.

After our discussions we were given a tour of his lab by his senior assistant, a man who had received his PhD some years earlier under his supervision. Any reference made by this individual to Dr. Villa was prefaced by the honorific *maestro*, and allusion to his work was reverent, bringing Browning's *A Grammarian's Funeral* to mind.

I have since found this old-world veneration by former students for their chief mentors to be the norm not only in Mexico but in most other countries, especially in Europe. At this time, I had had about a year and a half of experience with the traditions and philosophy of postgraduate education in the United States and had had no similar exposure to a maestro system. In fact, it seemed to me then, as now, that the tradition of student-mentor relationship in our country was quite different, involving a gradual maturation during which the student develops his own ideas, becoming increasingly independent of his advisor. As one who has since served as an advisor and major professor for a number of excellent students, I can tell you that the mentor is not the likely source for the variance between systems. It is much more pleasant to be revered than questioned by persons whose knowledge of the field, at least as it pertains to their own research, is likely to be more current and greater in depth than your own. My explanation for the difference between the graduate system in the United States as opposed to the Old World is that it lies in the degree to which the mentor has control over the professional future of his students. In many countries, perhaps most, there are few positions, publishing venues, or funding sources, and those that are available are tightly controlled by the senior people in the field. The situation is quite different in the United States because of the sheer size of the market. Famous senior mentors certainly can influence the careers of their top intellectual progeny: Jonas Salk comes to mind, who required that he have first authorship on any paper published by a member of his institute. Nevertheless, most US professors cannot control their students' development because the opportunities for professional advancement are too numerous. Whatever the cause, I believe the US system to be superior. Although it can be uncomfortable for mentors to be grilled and surpassed by their students, science is the beneficiary. Much more rapid progress in thinking and the destruction of superannuated paradigms are the results.

Our third critical interview was with Dr. Arturo Gomez-Pompa, renowned expert on tropical forest, with a seminal publication in the journal *Science* on the topic, and head of UNAM's Instituto de Biología. It was he who provided us with the solution to our study-site problem. He said he had no idea how we could do a multisite elevational sampling transect (sounds crazy to me too, in retrospect), but if it was tropical forest that we were after, then the best and closest place to find any amount of it would be in the Tuxtla Mountains of southern Veracruz state. Here about a third of the original rainforest remained, including a tract three and a half miles long and half a mile wide on the slopes of Volcán San Martín, where UNAM had a field station. He said that if we went there and found it suitable, we would be welcome to use it as a study site as guests of the Instituto.

We left Mexico City on February 16 and proceeded east across the plateau and down the Sierra Madre toward Veracruz city, a route much the same as that followed by Cortés and his merry band (in reverse) on their first visit to Moctezuma, and one traveled by Dwain many times with many students.

As we neared the great port, he reminisced about his first graduate student at Minnesota, Byron Harrell. The late Dr. Harrell (he passed away in 2010) was the type of student Dwain liked best—the kind with whom one could sit down, figure out a project, and find the funding. Then the student left, coming back in a couple of years with the data.

Dwain recalled heading north from Veracruz city with a couple of students along the coast in the early 1950s, some decades before there was a paved road. In fact, there was no road—just a series of dusty cattle tracks connecting villages. Thirty miles from anywhere, they encountered a jeep, which Warner recognized. Dwain had presumed Byron was somewhere in eastern Mexico, but neither he nor his family had heard from him in some months. He got out and checked the jeep. "Hah!" he said. "A beer says we meet him carrying a gas can a couple of miles up the track." And so it was. They did, in fact, meet Byron shortly thereafter, who greeted them laconically, politely resisted offers of assistance, and then continued on his way.

That night we stayed at the Mocambo in Boca del Río just south of Veracruz city, a beautiful old hotel that was probably an upscale bordello back in the 1930s, with swimming pools, terraces, and gardens for the comfort of guests not otherwise occupied. Prices and clientele, as represented by our bedraggled selves, had come down quite a bit since its heyday, but it was comfortable enough.

The next day, we headed south, crossing the Papaloapan River at Alvarado and its vast Lerma marshes. These wetlands, with their stunning numbers and varieties of waterfowl, herons, and shorebirds, are storied among field biologists. Former Smithsonian secretary Alexander Wetmore and his famous field man, Carricker, had worked there in the 1940s, as had many other cynosures in the tropical biologist list of luminaries. Of course, Warner had worked there, as had his most famous and favorite graduate student, Bob Dickerman. (This was also where Bob had shot a man, accidentally, while collecting rails in the marshes. As Bob explained, he only winged him, with #9 shot, and ten dollars seemed to make all well.)

Warner loved to tell Dickerman stories, including how they had met in the wilds of western Mexico when Bob was working as a professional collector for the great University of Kansas mammalogist E. Raymond Hall. But the best Dickerman story for my money is now the stuff of legend. When Dwain finally convinced him to pursue his doctorate at Minnesota, Bob traveled up to the Twin Cities to do his coursework and take his doctoral preliminary oral exams ("prelims"). It was during this interlude that he was driving down a Hennepin County backroad when he saw a thirteen-lined ground squirrel (a.k.a., "Minnesota gopher"), run out into the middle of the road and copulate with a dead female lying curled sideways in a manner similar, presumably, to a hibernating individual. Bob thought this interesting. Male ground squirrels normally copulate with sleeping females during hibernation, but still. In any event, he wrote up the observation and published it as a note in the prestigious *Journal of Mammalogy*, the leading journal in the field (Dickerman 1960). He titled his paper "Davian Behavior Complex in Ground Squirrels." None of the reviewers or the editor inquired as to what *Davian* meant, and the paper has been cited many times since—Davian behavior in mockingbirds, Davian behavior in raccoons, Davian behavior in moles, bugs, slugs, and so on. Most recently I found it in Jennifer Ackerman's new book *The Bird Way*, wherein she goes on at some length regarding the many bird species in which this disgusting activity has been observed, along with some pretty fancy psychological speculation on causes—still no explanation of *Davian*, though.

Now the story can be told. Dwain, as well as most of his students and colleagues, loved bawdy stories and dirty limericks. One of his favorites was the following:

> There once was an old hermit named Dave
> who kept a dead whore in his cave.
> He said I'll admit

I'm a bit of a shit.

But think of the money I save!

South of the marshes, we headed toward the Tuxtlas, their towering scarps capped with clouds looming ahead. Though located only sixty miles or so from Veracruz city, the Tuxtlas are remote. The reason for this is that they have some difficult terrain of steep cliffs, deep gorges, and cataracts along with rainfall levels that in some places exceed twelve feet per year.

In those days, to get to the University of Mexico's field station, you had to drive right through the heart of the Tuxtlas to the lovely lakeside town of Catemaco and then follow a winding, muddy track twenty miles northward, past the Coyame turnoff, over the Dos Amates pass, and down through the lowland villages of Sontecomapan and La Palma, crossing the usual split-log bridges, until you came finally to a rusted sign that constituted the sole evidence of the presence of the University of Mexico at the site. And forest. Much of the forest was gone in the Tuxtlas, even by 1973, but not at the biology station. There, ancient rainforest still loomed over the temporary watercourse that passed for a road.

Figure 4-8. Selva understory near UNAM's *estación de biología*.

On February 20, we finally got to the UNAM field station and had our first good look at primeval tropical rainforest (*selva* in Spanish), a truly spectacular environment containing all manner of exotic flora and fauna, with huge, buttressed forest giants hung with lianas and bromeliads through which howler monkeys and sloths skulked and parrots and toucans shrieked.

Field station turned out to be something of a euphemism for the two dilapidated guard huts we found at the site, but UNAM's patch of *selva* was nearly untouched. When we saw the forest, we knew we had found the site for our study. Our only problem then was to work out the logistics, beginning with where I could find a place for my family to stay. The guard said that should be no problem. Go back south two miles, and we would see a sign for the *cabañas* of Hotel Playa Escondida. Simply follow that turnoff to the end, and we would come to the promised hotel.

When we arrived at the "hotel," I remember that it wasn't raining. But it had been, and it would later. In fact, it is hard for me to think of the Tuxtlas without thinking of rain. It rains there all the time, although less so in April and May. On driving into the compound, which was located on a spectacular bluff 150 feet above the ocean, we met the proprietor and his wife, Raul and Julieta Garcia, and the only two paying customers, a rather sad-looking young Harvard professor who was studying tropical wasp ecology and his undergraduate assistant. The Harvard pair had had abysmal luck. Since their arrival a week earlier, it had rained continuously, and they had found only one nest of the proper species.

Despite their gloomy warnings regarding weather and work conditions, we were ecstatic. The forest was great. The birds were great. And there was even a place for us to stay while we were doing the fieldwork. What more could one ask? The only thing that bothered me was the long drive from the *cabañas* to the biology station. I could picture traveling the two miles on the (slightly) improved cattle path that led from the main road to the *cabañas* plus the two miles along the main road to the station—a half-hour drive when it wasn't raining—day after day, back and forth. It seemed like a waste of time and effort. Particularly since, from the topo map, the distance as the White Hawk flies was no more than half a mile. I was sure I could walk it in less time than the truck could drive it.

We had arrived in the early afternoon, and once settled in our concrete block rooms, we headed back over to the UNAM station to set up mist nets. We worked until about six or seven nets were up and then started back for the *cabañas*. It was about 4:30 in the afternoon. I had been expounding to my companions on how easy it should be to walk from the station to the *cabañas* because of the steep isosceles triangle formed by the main road and Raul's. I finally demanded to be let out so that I could prove it to them. I cut straight down a steep slope and across a field of dense second growth and in fact met the truck as it was slowly negotiating its way along the last half mile to the *cabañas*.

This demonstration, however, was not sufficient. I explained that there must be a path used by the local folks and that I would start at the *cabañas*, find it, proceed to the station, and then return. It was about five in the evening by this time. I set out heading straight west toward where I knew the station was located, following my compass. I went up, and up, and up through beautiful forest until I reached the top of the little reason for the detour followed by Raul's road: Cerro Balzapote. Cerro Balzapote is just under one thousand feet in elevation according to the surveyor's marker at its peak, and it lies exactly in between the biology station and the *cabañas*. There were no paths, except for those followed by mazama deer, peccaries, and an occasional hunter.

By the time I reached the top, light levels had gone well past the murky stage. Starting out half an hour before in the brightly sunlit pasture, I thought that I had until at least six before it got really dark. I found that in the forest, night comes early in February. By 5:30, the light was nearly gone, and I knew that I could not be even halfway to the station. Most smart folks would have turned around at this point, but I kept thinking that I would soon hit the road. And besides, it was mostly downhill.

As I set off down the western slope, night noises were beginning to build to their evening crescendo: katydids, frogs, bats—marvelous, unknown stuff. An owl hooted, *"Whetu whew whew whew."* I was thrilled. Everything was so new and different, and as the light disappeared completely, I felt as though I were inside the forest, like Jonah in the whale. By the time I was halfway down the hill, I heard another owl species. At least I suspected it to be an owl; it had a muffled, descending *"Bu bu bu bu bu bu bu"* call. It was also completely dark. So dark that the only things I could see were twigs and branches covered with brilliant green phosphorescent fungi and the glowing "eyes" of elater beetles.

Since I had no light, the compass was of no use. Nevertheless, I still thought I would hit the road soon and so kept stumbling along downhill. Instead, I came to a boulder-clogged arroyo. I found this discovery to be disconcerting. The creek did not fit with the lay of the land in my head. It forced me to doubt my sense of direction, so I finally had to accept that I probably could not find the road in the dark. Now a second worry began to weigh on me. My companions would surely be concerned. Even now they were probably busy trying to get Raul to form a search party from the nearby fishing village of

Jicacal. What an embarrassment that would be. So I decided to hurry along as best I could, hoping that they would wait at least a little bit before rousing the countryside.

I reasoned that a stream could end at only one place in this part of the country—namely, the ocean—and that once at the ocean, I could follow the beach to Raul's *cabañas*. The stream also had the advantage that one did not need a flashlight to follow it. The logic was impeccable, but the giant boulders along the arroyo made for tough going in the woolly dark.

After a half hour or so of groping, falling, and crawling over rocks and logs, I came out into the open. The trees were gone, replaced by a brilliant starlit sky. I could now see a bit—enough to know that I had come to the border of a pasture delimited by a barbed wire fence. I climbed through the fence and up out of the arroyo. Off to my left at a distance of five or six hundred yards, I saw what seemed to be car lights. Also, there were figures moving in the lights, and I could hear a horn honking. Now I was quite sure that this was the search party. Thoroughly chagrined, I started walking off across the pasture toward the lights.

After proceeding just a few yards on this route, I heard a heavy galumphing sound and glimpsed a large white form heading rapidly in my direction, causing me to sharply increase my pace. Just moments earlier my fatigue, along with the bumps and bruises sustained in my various tumbles, had made me hard pressed to move at all. But with the approach of that form, which I took to be a Brahma bull, I found myself fairly flying over the grass. Still, the bull was faster. Fortunately, I had a head start and reached and dove through the fence on the far side of the pasture a few yards ahead of my pursuer, leaving only a few layers of skin on the barbs in the process.

I picked myself up and found myself in a stubble cornfield. The vehicle lights had disappeared. Depression. My rescuers were already heading off, I supposed, spreading out in all likely directions to find me. Oh well, best keep moving toward what must be a road, in any case. A few minutes later, I saw a tiny light come bobbing up over a rise and move along diagonally away from me only thirty or forty yards ahead. I ran up to it and found a little girl, seven or eight years old, walking through the stubble carrying an oil can with a lighted wick. I was conscious of the fact that I must appear as some sort of terrifying apparition, looming up out of the dark, but I didn't know what else to do.

"Hola," I said.

"Hola, señor. ¿De dónde viene usted?" she said, sensibly enough. Unfortunately, all I understood was the *hola*. As I have said, my training in Spanish began on the first day of our trip, when Warner made me go arrange for rooms for the night. So in response to her polite question, I said, "No hablar [*sic*] español."

"Veng," she said, and since I did not respond to her request, she reached out and took my hand and started to lead me along like the helpless idiot I was. We walked a hundred yards or so and then, to my great surprise, entered a rough track bordered by huts. We were in a village, but there was practically no light whatsoever, only occasional flickers of firelight seen through cracks and spaces in the stick walls of the huts. I hadn't even seen the place until we entered it. We walked together a bit farther and then rounded a corner into a pool of light from the only bulb in the place—hanging over the front door of a hut that served as the village store and bar. By this time, we had attracted a crowd of twenty-five or thirty people of all ages and sexes.

At this time in my career, I knew the words *cerveza*, *quarto*, *baño*, *hablar*, *camino*, *cabañas*, and *biólogo*, and I have to say that the prospect of trying to convey my predicament and learn the way back to the hotel using this vocabulary in my fatigued state seemed daunting. My audience, however, was undeterred. They thought I was great fun. They offered me food and drink while asking me a hundred questions, to most of which I could only smile foolishly and say "No hablar español." Nevertheless, I eventually understood that I was in the *ejido* Balzapote.

They offered perfectly logical solutions to my problem. First, they told me I was quite welcome to stay the night; a bed would be found with no difficulty whatsoever. They made me understand that if I chose that option, there would be a truck coming by in the morning that could take me right to Playa Escondida. When they understood that I really wanted to get back (still worrying about the search party), they said to take a horse. I told them I didn't think that was a good idea. I just wanted to be put on the *camino* to the *cabañas*. That turned out to be easy enough. Like Dorothy in Munchkinland, I was standing on it. I just had to keep walking along this road until I came to Raul's sign, then walk on in—about a five-mile stroll, but no rocks, logs, or Brahma bulls. Piece of cake.

Before leaving, I made one request: Could they please sell me a lantern like my little guide's? They tried to present it to me as a gift, but I gave them ten pesos for it, about $1.50 at the time. I still have that lantern.

We said our goodbyes, shaking hands all around, and I headed down the road. This was my first experience with poor country folk, but I have found it to be typical in many other places visited since. They tend to accept people as they are, until you show them differently, and they are the warmest, best people in the world.

The walk back was long but completely uneventful. I arrived at the *cabañas* at about nine that evening. As I limped up the road, I could see my companions seated around a table on the restaurant patio, drinking, laughing, and joking. There was certainly no evidence of the imminent organization of a search party or any other signs of distress.

As I stepped into the light, Dwain said without the slightest indication of surprise at my bedraggled appearance or tardiness, "Hello, John. Find anything interesting?"

"Sorry I'm late," I said.

"No problem. Unfortunately, we couldn't hold dinner, but there's plenty of beer."

"I was in a panic," I said. "I thought that you all would be worried. Maybe down at Jicacal organizing a search party or something."

"Well, no. We hadn't thought much about it. Figured you'd probably got turned around a bit at night in the woods and would show up by morning. I suppose if you hadn't appeared by tomorrow night, we would've started to think about where to look for you, although that's a mighty big place out there. Probably wouldn't have made much sense."

They were interested in the two owls that I had heard. One was the Mottled Wood-owl, *Ciccaba virgata*, a common species in the region, but the other they thought might be a Spectacled Owl, *Pulsatrix perspicillata*, for which there were no records at the time in the Tuxtlas (since found to be fairly common where lowland forest remains). After that, the talk drifted to other subjects and plans for the next day's work.

So I learned a few lessons. First, a good field biologist does not go out looking for problems. Second, if he is so foolish as to do so, it's his own affair, and he shouldn't expect others to waste their valuable time on his stupid mistakes regardless of the amount of style or dash. Third, it is the quality of the information that one brings back that is of primary importance, not the incidental occurrences met along the way. I took those lessons to heart, though I cannot say that I have always acted on them.

And I still think that they could have pretended to be a little worried (Paul has since noted that *he*, at least, *was* concerned—but then he was still just a field biologist in training).

Since Mexico's people, politics, and culture as well as its natural history play critical roles in the story of my next few decades of research in the country, I hope you will have patience while I digress here and there on these topics. This particular digression has to do with Mexico's land tenure system. The village into which I had stumbled on that first night in the Tuxtlas was Ejido Balzapote. An *ejido* is a form of land tenure instituted after the Mexican Revolution (1910–20), an apocalyptic holocaust during which one and a half million people died and a quarter million fled (mostly to the United States) out of a population of fifteen million. *¡Tierra y libertad!* was a rallying cry and land reform a major issue, and in 1934, the government of President Lazaro Cardenas passed the agrarian code, which redistributed over forty-five million acres to landless peasants (*campesinos*). Much of this redistribution was in the form of *ejidos*, a sort of commune created by petition to the federal government by a group of *campesinos* for a land grant. Under the terms of the grant, each adult male member of the commune (*ejidotario*) has usage rights in perpetuity to a portion of the land, which he can pass on to his children. However, the state owns the land. The *ejidotario* cannot sell it, and if he fails to use it for more than two years, usage rights return to the *ejido*.

In addition to the *ejido*, there is another type of land tenure that figures in my story—namely, that of the *colonia*. Not to be confused with the colonias of South Texas, which are unincorporated collections of squatters, a *colonia* is another method for distribution of government land. The difference from the *ejido* is that the purchaser owns the land, twenty-five hectares (about sixty acres), and can do with it whatever he wants. Most land tenure in the Tuxtlas is in the form of *colonias*, and most of them are operated as ranches. Economically, *colonia* owners represent a big step up from *ejidotarios*. Many were absentee, living in Catemaco, and hiring *ejidotarios* or Popolucas (indigenous Indians of the Tuxtlas) to do any work needed on the property. Raul and Julieta's property likely was a *colonia* of this type.

The next week was spent doing fieldwork in the forest at the station, mostly in the form of mist-netting. We set up about ten nets and checked them several times a day to remove what was in them. Most birds captured in this way were collected (killed) and prepared as museum specimens for the Bell Museum bird collection.

Although our time at Playa was brief, I made a couple of observations that were important in shaping my planned studies. First, wintering migrants could be found in apparently undisturbed *selva* and cloud forest, contrary to much published literature on the topic. Second, there wasn't much left of either of these habitats in the entire Mexican state of Veracruz.

CHAPTER 5

PREPARATIONS, DEPARTURE, AND EARLY TEXAS EPIPHANIES

I am the wrongness of here.

—LES MURRAY, "MIGRATORY"

By mid-March 1973, we were back in the Twin Cities, where I began to build the strategic plan for my thesis fieldwork, set to launch in the fall. I spent those months prior to our departure for Texas immersed in the literature of migratory bird nonbreeding-season ecology, most of which interpreted the migrant during this period away from the breeding ground as subsisting in the classic "wanderer" mode of existence. The prevailing theory regarding migrant ecology and evolution at that time (and now) was as follows: The migratory birds of temperate and boreal North America originated (evolved) as resident birds in these regions during geologic periods when the climate remained equable for their survival throughout the year. As the climate became more seasonal, with increasingly harsh weather during the nonbreeding seasons, these temperate zone residents gradually evolved the various attributes required for successful migration, such as the accumulation of fat stores preliminary to movement and navigational capabilities enabling them to move southward as the weather deteriorated and return to their breeding areas as it improved. Naturally, southward movement from their breeding areas brought migrants into regions where members of other bird species lived throughout the year. These resident birds filled all of the available niches (ecological space) in the stable habitats of these regions, forcing the invading migrants to subsist as wanderers throughout the migratory and wintering periods, depending on superabundant foods in marginal habitats that could not be completely harvested or defended by the resident bird community members.

The hypothetical problem confronting migrants regarding the absence of a niche on departure from their breeding areas had been clarified in the 1960s by the theoretical work of Princeton's Robert MacArthur and his colleagues. He explained that south of migrant breeding areas, the niches held by them were filled by other species, often close relatives, that were nonmigratory. These residents he deemed "ecological counterparts" to their migratory brethren, theorizing that they would prevent transients or wintering birds from occupying niches in stable environments—either directly by attacking them or indirectly by out-competing them—thereby relegating them to wanderer status dependent on temporary food concentrations.

This understanding of migration, called the northern home theory, has been around for thousands of years. Homer mentions it; the Bible mentions it; Aristotle mentions it; Pliny the Elder mentions it. And it is still the idea of how migration works taught in most college classrooms today, albeit with considerable elucidation and elaboration from such rapidly changing fields as physiology and population genetics.

It was this conception of the principal evolutionary outline governing migrant ecology that shaped the plan for my study of the phenomenon of migration, which was as follows:

Fall Period: August–October 1973 at the Welder Wildlife Refuge on the South Texas coast investigating ecology and behavior of migrants, mostly songbirds, as transients

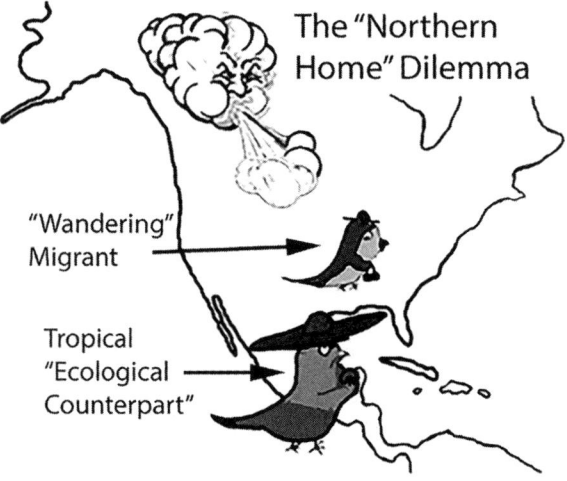

Figure 5-1. "No queremos su tipo aquí!"

Winter Period: November 1973–mid-March 1974 investigating ecology and behavior of migrants on their wintering grounds in rainforest and second growth of the Tuxtla Mountains of southern Veracruz

Spring Period: mid-March–May 1974 back at Welder studying spring migration

The second year of work, August 1974–May 1975, would follow a similar breakdown of activities.

Of course, I was still a graduate student performing daily duties as a curator in the bird section of the Bell Museum and, commencing the spring quarter, serving as Warner's teaching assistant in ornithology and taking a class in mammalogy from Elmer Birney.

Elmer was a young assistant professor and curator of mammals at the museum at the time. Like me, he had been in the service—the navy, in fact—and had then gone on, like Bernardo Villa, to complete his PhD at Kansas under E. Raymond Hall. It was Elmer who taught me an important and necessary lesson for success as a field biology professional. It came via a paper that I had written for his class—on field mouse (*Microtus*) ecology, as I recall. Therein I was holding forth on the "value to the species" of some odd behavior. Instead of simply giving me an F, which is what most ecology faculty would have done, he called me in to his office, sat me down, and explained the intellectual problem posed by this phrase.

"Do you know who Wynne-Edwards is?" he inquired, to which I answered in the negative. From there he went on to explain that Wynne-Edwards had written a book in 1962, *Animal Dispersion*, famous among behavioral ecologists, in which he presented exhaustive evidence from field observation supporting the theory of group selection, which holds that some kinds of animal behavior can be explained only by their value to the group (tribe, population, species) to which the individual belongs, since the value to the individual (in terms of the genetic contribution to the next generation, which is the ultimate basis for natural selection) is not obvious. Elmer further explained that population genetics theory regarding the operation of natural selection provided no theoretical support whatsoever for group selection and that nearly every example provided in the book could be explained on the basis of individual selection, as pointed out in a large number of published papers. The only possible exception was in the case of apparently altruistic behavior among closely related individuals, which W. D. Hamilton had dubbed "kin selection."

"So you see," he said in the kindest way possible, "using the term *group selection* or stating or implying that a behavior results from anything other than individual fitness [that is, the individual's genetic contribution to the next generation] is evidence of a profound misunderstanding of how natural selection works. These elementary principles probably were not explained clearly to you in the army, so I am going to give you the opportunity to rewrite this paper incorporating your newfound comprehension."

I relate this anecdote because it has been my experience that people who are not professional biologists often interpret animal behaviors based on their understanding of what is best for the species. This approach is natural for humans. After all, the fundamental tenet of the Christian religion, "Do unto others as you would have others do unto you," is about as clear an expression of altruism as one can find. However, our species is different from others, and if you attempt to view behaviors of other species through the lens of experience with your own, you are unlikely to progress far toward understanding their ecology and evolution. Put another way, the only path to understanding migration is to consider it from the perspective of what is best for the individual bird. Without this viewpoint, many of the arguments presented in the following pages will be incomprehensible.

Back in 1973, the University of Minnesota Field Biology program offered intensive courses during two six-week summer quarters at a rustic campus on the shores of Lake

Figure 5-2. Dwain Warner and Dick Oehlenschlager preparing for the Itasca pig roast, July 1973, with my daughter, Brigetta, in bemused attendance.

Itasca in northern Minnesota. For the first summer quarter of 1973, Bonnie and I with the kids moved up there, where we lived in three tents while I served as a teaching assistant in Warner's field ornithology class and took a class in field botany.

It was during the second or third week of our stay that Warner brought Mario Ramos to our campsite and introduced him as the Mexican student who would be joining our team on the Welder grant. Mario came highly recommended by Allan Phillips as the top student in ornithology at UNAM. In Mexico, the baccalaureate degree is more similar to a master's in that it requires a senior thesis. Mario had completed his on the birds of the Pedregal, a region of volcanic ash and solidified lava flows located outside Mexico City, with Phillips serving as major professor. Mario became my closest professional colleague, and he and his family among our closest personal friends. We worked together on many different projects over the next thirty-two years until his untimely death in 2006.

On completion of the summer session at Itasca, my family and I sublet our apartment in Saint Paul and took off south across the Great Plains of southern Minnesota, Iowa, Missouri, Kansas, Oklahoma, and Texas to Welder on the western Gulf Coast. We arrived there August 1 after a three-day drive.

During our earlier trip to Welder, we had stayed in one of the married-student apartments in the dormitory. However, we had found this setup a bit tight for us and our two kids, especially with the rest of the dorm filled with students. Fortunately, we had found out that the little red house near Moody Creek, a mile or so from the campus, was going to be available in the fall and had been given permission by Dr. Cottam to land there.

Welder was then, and is now, an exceptional organization for field biology studies. It was established by a statement in rancher Rob Welder's will, which required simply that a foundation be set up "to further the education of the people of Texas and elsewhere in wildlife conservation." This idea was translated into reality by the brilliant decision of the trustees to hire Dr. Clarence Cottam as director. Dr. Cottam was one of the preeminent wildlife biologists of his day. As a scientist in the US Fish and Wildlife Service (USFWS), he was the first (1946) to publish a study on the deleterious effects of DDT on wetlands ecosystems, and as the assistant director of that organization, a promoter of the work of Rachel Carson, cited by her in the acknowledgments of her book *Silent Spring*. The Welder trustees hired him away from Brigham Young University, where he was dean of the College of Agriculture (and a bishop of the Mormon Church). It was Cottam who translated Rob Welder's rather vague vision into reality—graduate student heaven.

Dr. Cottam ran the Welder Foundation as a first-rate research facility, treating the graduate fellows like the preeminent scientists he hoped they would become. Like Dwain, he loved a good story. One of his favorites dated from when he was the assistant director of the USFWS. He and a colleague were deputized by the USFWS director to travel around the country holding hearings at major duck-hunting areas to explain and defend the "take" laws the federal government was putting in place to be sure that populations were maintained at healthy levels. One such place was at a large gathering in an auditorium in Seattle. Dr. Cottam would tell it like this: "Now, these meetings were often contentious. Lots of people hate the government, and they especially hate being told by that government what they can and cannot do, and that was the case in Seattle. There was a strong sense of outrage among duck hunters there regarding the small number of each duck species that a hunter was allowed to shoot. The president of this group was a lawyer when he wasn't duck hunting and a skilled advocate for them. My colleague was leading the presentation, explaining the reasoning for the bag limits, when he was rudely interrupted by the duck society president. 'Isn't it true that the bag limit for pintails is set way too low?' he asked."

Here, Dr. Cottam would say, "My colleague should have answered simply 'That's not true.' But he didn't. What he said was 'I don't think that's true, sir,' to which the lawyer said, 'Well, is it true or isn't it?' Now here my colleague really caused a problem by saying 'We believe it to be true, but we will do studies to make sure.' Well, that is exactly what the lawyer wanted to hear. He turned to the audience and said, 'There you go. The government is always happy to tell us how to hunt our ducks, and if we want to know why, they say they will use our tax money to do another useless study. Studies, studies, studies.'"

At this point Dr. Cottam would say that he got to his feet on the stage to help his colleague respond, saying, "Sometimes there are not clear yes or no answers to complex questions, sir." To which the lawyer responded, "That's not true. All legitimate questions can be answered yes or no." To which Dr. Cottam said, "That's simply not true." To which the lawyer replied, "You cannot give me one single legitimate question that does not have a yes or no answer." To which Dr. Cottam replied, "OK. Here is an example: Are you going to stop embezzling the funds of this organization?" Chortling, Dr. Cottam would describe the apoplectic lawyer opening his mouth in preparation for a thunderous retort, pausing for a moment, and then abruptly sitting down, never to make another peep, thereby drawing the hearing to a successful close.

I probably heard this story about ten times in the short year we had with him.

The Welder Wildlife Refuge is seven square miles of native thorn forest, oak savanna, riparian woodlands, marsh, and pasture located along the south bank of the Aransas River about ten miles inland from Copano Bay. The campus is composed of several beautiful structures in Mexican colonial style, including an administration building, library, laboratory, museum, study, lecture hall, and student and senior staff housing.

In addition to Dr. Cottam, the staff at the time was composed of Caleb Glazener (assistant director) and his wife, Willa (librarian); Gene Blacklock (museum curator and director of education); Mr. Maley (ranch foreman); Celso Villareal (ranch hand); Rosemary Covington (secretary); newly arrived assistant director Eric Bolen and his wife, Becky; janitorial staff, Mr. and Mrs. Garza; and three landscaping workers.

By August 3 we had gotten ourselves into our new digs, found a day-care person for the kids, and begun the task of setting up our study site. During our visit of the previous winter, I had roamed the refuge, searching for the best place for capturing large numbers of transient birds, and had chosen Hackberry Mott. The mott (grove of trees) was a strip of riparian forest about half a mile long and three to four hundred yards wide located along the banks of the Aransas River just a mile or so east of the campus. Hackberry, cedar elm, pecan, and anacua (subtropical understory shrub) were the dominant

Figure 5-3. Bon at the north end of Hackberry Mott, August 1973.

Figure 5-4. Bonnie and fellow Welder Fellow Joe Folse (studying roadrunners) helping me set mist nets in Hackberry Mott and evidencing a distinct lack of respect for the principal investigator, August 10, 1973.

trees. Within two days of our arrival, I began cutting lanes through the dense, thorny thickets of the mott, readying them for net placement.

Individuals of some migrant species were already passing through on southbound migration by the time we arrived at Welder, including the Kentucky Warbler, Louisiana Waterthrush, and Yellow-bellied Flycatcher. The observation and capture of these birds illustrated a major attraction of Welder as a field site for the study of migrants—namely, its location along the southern Gulf Coast of Texas, where the vast majority of birds occur in passage only, neither breeding nor wintering in the region. This attribute means that one can be sure that the behavior and physiology of these birds are that of transients, an important aspect in trying to determine key characteristics of migrants on their way to and from their breeding areas. We knew that the birds we saw and captured were on the move.

Like Ibn Alhazen, Francis Bacon, and indeed all good empiricists, Pasteur emphasized the importance of the experimental method in determining the validity of any given hypothesis. For the field biologist, this presents two large problems that the laboratory researcher does not have: first, the need for a clear understanding of the assumptions that you are making in your experimental design and analyses, which can in fact take years to develop; and second, a way to separate noise from signal, by which I mean how to distinguish the important data from the immense amount of information coming in that bears little or no relationship to the questions at hand. Confronted with these issues, the field biologist must continually guard against drawing conclusions based on inductive reasoning from one or a few observations. The inductive process is fine for hypothesis formulation but of little value for building a solid understanding of natural processes. Such comprehension comes best through the standard application of the scientific method. However, it is not always possible to follow deductive procedures because of the difficulty in constructing a meaningful test where all assumptions are recognized and controlled; however, the disciplines of continually trying to come up with such a test and recognizing the limitations of your inductions when you fail are important to keep in mind.

Hypothesis formulation and testing are key parts of this issue of attempting to sort observations into meaningful explanatory constructs. Ronald Fisher, who laid much of the foundation for statistics, proposed that the base, or "null," hypothesis should state the assumed situation that there is no relationship between two phenomena. A test is formulated on this basis, data are collected and subjected to a statistical analysis, and a probability is calculated for the likelihood that the null hypothesis is correct. If it is found to be incorrect, usually at a probability greater than 95 percent, then it is rejected, and a new, alternative hypothesis is formulated. That is the reasoning that I have tried to follow in my own research and present in this account. It seems backward, of course. Shouldn't you just state what you think you are going to find and not play any mental games? That approach is what they now expect in National Science Foundation (NSF) proposals, or at least that was the case in the early 2000s. My own sense is that by stating what you think you are going to find, you lose some objectivity, which is always a precious commodity for a research scientist. With these ideas in mind, we began our first field season.

Once the net lanes had been cleared, I began putting the mist nets in place with help from Bonnie and fellow Welder student Joe Folse, placing fifty mist nets with two nets set about every hundred feet or so perpendicular to the lane traversing the mott from north to south.

Welder is located along the main migration pathway in North America. Millions of individuals of over three hundred species of birds make semiannual peregrinations along this corridor, and at times it felt like most of them were in our nets. Things started

Figure 5-5. One of Bon's little "friends" (a cottonmouth), which she accidentally stepped on during an early morning net check. She always wore her snake leggings, so no problem.

fairly slowly in mid-August, when we first began netting, but by late August, we were averaging more than fifty captures a day that often included ten to twenty different species. Bonnie was helping me with both checking the nets and processing the birds, but the load quickly exceeded our ability to keep up, forcing decisions regarding what was most important in terms of understanding migration.

My first effort to reduce the load was to release without recording or banding all permanent residents, such as cardinals and Carolina Wrens, and summer residents, like the Painted Bunting, Yellow-billed Cuckoo, and White-eyed Vireo. This decision helped, as these species had constituted a significant portion of our captures, but we were still overwhelmed, not only by actual work, but by the sheer volume of information coming in. What was I to make of the data we were collecting in terms of understanding transient biology, and was our methodology the best use of our time in the field?

These questions were key, and it was important to answer them as quickly as possible; indeed, right now as we were setting the course for the next two years of fieldwork in Texas and Mexico (and Minnesota, as it turned out). The research idea that we had written up while conceptualizing in Warner's Bell Museum lab back in the fall of 1972 focused on an assessment of differences in the timing of passage movement by age and sex for different species and subspecies in Wisconsin, based on TV tower specimen data, and the Texas and Veracruz Gulf coasts, based on netting data. It was a good plan, theoretically, but as is often the case, it was fundamentally flawed.

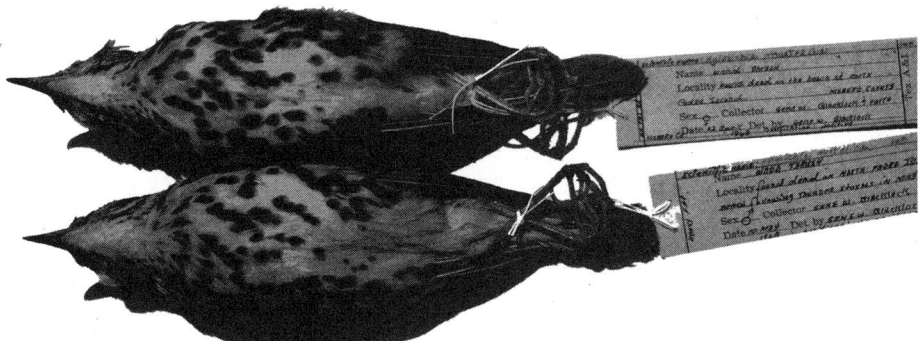

Figure 5-6. Museum skins (specimens) of the Wood Thrush.

The first problem with our design was that the data collection procedures would involve an immense amount of work with dead birds because you pretty much have to have the specimen to be able to sex and age members of most species reliably during the nonbreeding period. Also, the identification of a bird to subspecies normally requires a comparison with large samples of museum specimens representing the different groups (see figure 4-7). As a result my time in the field in both Texas and Mexico, had it gone according to plan, should have involved capturing migrants in mist nets; killing captures of those species of particular interest (that is, those represented in samples from both Wisconsin and Veracruz); recording sex, age, fat, and molt for each specimen collected along with critical measurements (usually wing, tail, tarsus, and bill); and then preparing it as a museum skin. These procedures were, in fact, those that were being followed religiously by my colleagues Oehlenschlager and Mario down in Veracruz at the same time that I was working in Texas.

The second problem was that time spent capturing birds and preparing specimens was time *not* spent observing what migrants were actually doing. As it happened, my activities during the first days after our arrival placed me perfectly for observing transient behavior. The work of cutting fifty net lanes in my riparian woodland study site put me in the same environment as the migrants for several hours a day over an eleven-day period. During that time, we (the migrants and me) were all in the same steamy scrub together (it was a hot and rainy August), and while I was working, I was watching and listening to them, thinking about what I had read of their lives as transients. Here in Hackberry Mott, they were no longer data points or theoretical constructs. They were individuals doing their best to stay alive long enough to get back up north to reproduce.

Their situation became very real to me, and what I saw and heard during those few days changed the focus of my investigation profoundly. I was well aware of the importance of the questions that our proposed work could answer regarding the structure of migratory movement. However, they were the kinds of questions that could be answered by anybody at any time who had access to the specimens. You (the researcher) did not need to be the one doing the collecting. The more time I spent with the birds in Hackberry Mott, the more I came to realize what an unusual gift I had been given. Through the generosity of the Welder Wildlife Foundation, I had two years during which I could spend whatever time I wanted with migrants in the field. It seemed to me that not using as much of that time as possible watching and listening to what the birds were doing represented a waste of an incredible opportunity. For me it appeared that time spent working with dead birds while in the field was time stolen from watching the live ones.

In addition to these thoughts, by the time we arrived at Welder, I was already skeptical of aspects of the northern home theory—according to which migrants in winter should be homeless visitors to marginal habitats, such as hedgerows and overgrown pasture—because of observations made in February in southern Veracruz, which seemed to indicate that some members of some of the migrant species wintering there were living in untouched primeval rainforest. Therefore, I began my fieldwork in South Texas looking for behaviors that might not fit the northern home paradigm, and behaviors are exactly what dead birds can't provide. While continuing to net, I shifted the emphasis on data collection from dead birds to live ones, banding and releasing most of our captures. This shift in emphasis quickly lead to three observations from our first weeks of sampling that drew my particular interest.

The first observation was that the earliest migrants (August) to arrive in South Texas were *Empidonax* flycatchers (Yellow-bellied, Willow, Alder, Acadian, and Least) and birds whose main breeding populations are located in the southeastern United States, such as the Kentucky Warbler and Louisiana Waterthrush.

The fact that *Empidonax* were among the earliest transient arrivals was not a surprise. Unlike most birds, the adults of these species migrate before undergoing the postbreeding molt, and so begin their southward movement some weeks before the majority of songbirds, which attempt to complete molt before departure. The early migration by southeastern breeding birds was more intriguing. If southward migration is a result of cold weather, as posited by the northern home theory, why should south-temperate breeding birds depart before north-temperate or boreal breeding species, and why should members of either group head south in August, when cold weather is still a couple of months away?

The second interesting observation was that recaptures of banded transients averaged less than 5 percent of total captures among the thirty-three migrant species recorded. This finding indicated that the vast majority of songbird migrants arriving at Hackberry Mott in the early morning after a night of migration departed by the evening of that same day. Only a small portion remained at the mott for more than twelve hours.

The third finding was that while most songbird migrants observed during the day were quiet, active, and social, foraging in loose flocks, a small portion, including some members of the same species mentioned as occurring in flocks above, were not social. Rather, they occurred as solitary individuals and occasionally vocalized using a chip note or, in some species, song. These birds often remained in the mott for more than one day. Franz Groebbels suggested, based on caged bird studies done in the late 1920s, that migrants alternated between at least two different physiological states over the course of their migratory trip: *Zugdisposition*, or a "feeding" state, in which they were preparing for migratory flight by eating intensively (hyperphagia) and laying down fat reserves, and *Zugstimmung*, or a "flying" state, during which they were ready for or undertaking actual migratory flight. It seemed to me that the birds in the different behavioral groups (social versus solitary) might signify these different physiological states, with the active, social individuals that stayed in the mott for only ten or twelve hours representing birds in a flying state and birds that were solitary, vocalizing, and staying more than a single day representing birds in a feeding state. I further reasoned that while birds in a flying state presumably would not care what sort of habitat they spent their few hours on the ground in or with whom, birds in a feeding state, attempting to build fat reserves, might be expected to behave quite differently—searching for optimal feeding habitat and per- haps even defending a territory against members of the same species (conspecifics).

A territory is a place from which intruders are excluded by some combination of advertisement (such as song), threat (in the form of visual displays), and attack, the purpose of which is to sequester a necessary, defendable resource (such as mates, food, or nesting sites) for use by the defender alone. When we began our Texas work, there were few reports of territoriality in transients, and of these few, the majority were based on observations of shorebirds or hummingbirds, not songbirds. Nevertheless, what we had seen in Hackberry Mott was indicative of some sort of defense by some migrants. Individuals of twelve species of transients had been heard vocalizing by September 12, four of which used brief chirrs or chips and eight of which used song or call solely or in addition to chirrs or chips. All of these migrant species occur solely as transients at Welder, neither breeding nor wintering in the region.

The discovery of territorial defense in a transient was potentially of great importance in terms of both expanding our understanding of how the process of migration works and testing the northern home hypothesis. As mentioned above, migrants are assumed to be wanderers once they have left their breeding areas, subsisting on the leavings of resident birds in the environments through which they pass during the nonbreeding period. The possession of a territory, even temporarily, could indicate that the bird was not a wanderer and that it was capable of defending a piece of ground in a stable environment, not only from conspecifics, but presumably from any "ecological counterparts" that might be lurking about as well.

These preliminary findings stimulated me to devise an experiment that might test whether the small percentage of birds that remained in the mott for more than a day were depending on (and therefore defending) its resources to rebuild fat reserves. The first such experiment involved displacing birds from their points of capture. My reasoning was that if a bird is defending a piece of ground, that piece is important enough for it to go back to if displaced before fat reserve repletion has been completed. The null hypothesis for this experiment was as follows: "Transient birds on southward migration *do not* show fidelity to particular stopover sites."

From the time of our initiation of fieldwork on August 3, we had banded and released birds where they were captured in Hackberry Mott. However, as of August 16, I began to release at least some of the birds a mile or two *south* of their capture points in the same habitat type (riparian forest) in order to test the null hypothesis stated above. It seemed to me that if even one individual migrant of any species returned northward to its point of capture (a behavior known as "homing"), I could reject the null hypothesis, at least for that species. Our data already demonstrated that the vast majority of transients captured in our nets remained in Hackberry Mott for less than a day. Clearly, these birds would consider a more southern release point a head start on further southward migration. I also reasoned that if even a few migrants did show fidelity to stopover sites, the occurrence would be indicative of possible territoriality, at least in those individuals.

The odds against any return seemed huge, particularly in light of the effort required in displacing the birds. Even if a small percentage of some birds of some species did stay for more than a day and did defend a territory, what would be the likelihood that we would recapture them after giving them a new place in new suitable habitat farther south along their intended route? Minuscule at best.

For the first couple of days, the expected zero results ensued. However, on August 19, I was checking mist nets along the margins of a shallow oxbow pond at the south end of the mott in a torrential thunderstorm when I found a Louisiana Waterthrush in the bottom shelf of net 46. To my amazement, it was a banded bird captured three days previously in the same net and subsequently released at Moody Camp, a mile southwest of the capture site.

This discovery was one of the great *Eureka!* moments of my career, and although I have had several since, some of which I will attempt to describe in these pages, none was quite like this. I literally whooped and leaped into the air when I saw the bird's band because I knew at that instant, in the midst of the intense fury of a Texas gully washer, that this little waterthrush represented something momentous. Clearly, the muddy strip bordering net 46 was important for this bird, Tex 88, an immature (born earlier in the year) of unknown sex. To solidify that point, it was captured three additional times in the vicinity: August 27, August 30, and September 5, returning each time after release a mile or more south of its capture point.

This stunning finding was indicative of at least three critical departures from the northern home theory:

1. This migrant was not wandering. It had a home, if only a temporary one.
2. Once it had established this home, it knew exactly how to find it again when displaced.
3. No resident "ecological counterpart" prevented its occupation of this home despite its presence in a stable, mature habitat (riparian forest).

However, there was little in the way of supporting data. Ultimately, I transported several hundred individuals representing twenty-five migrant species, only twelve of which (of six species) were recaptured near their original capture points. These results, while intriguing, were anything but conclusive.

Warner and I had the chance to discuss my research focus and methodological changes early in this first field season. Having finished helping Dick and Mario get set up and going at the UNAM field station in the Tuxtlas during July and August, Dwain headed back for Minnesota, stopping at Welder on September 5, having taken the bus all the way up from Veracruz, a thirty-six-hour ride.

He only stayed a day, but we had an intense discussion about how I should best employ my data collection efforts. He was keenly disappointed in the new directions I was taking and the methods I was using. He argued that keeping as many specimens as possible of those migrant species that had breeding populations whose origin could be identified by plumage (subspecies, that is) would be valuable to address the questions posed in the proposal regarding population movements. He further maintained that if I wanted to get a job as an academic museum curator of ornithology after graduation, experience with this type of taxonomic study would be invaluable. My argument was that while the differential timing of sex, age, and subspecies movements might both be interesting and involve excellent training, those questions were not especially relevant to the key issues as I understood them, such as why and how birds migrate.

He in turn was skeptical that I would be able to collect meaningful data relevant to addressing these issues: "You are making a serious mistake, John. A lot of smart people have been looking at the questions you mention for a good long while—brilliant scientists at the Grey Institute in Oxford, Germany's Max Planck Institute, and the Smithsonian, not to mention several senior ornithologists and their students at top universities here in the States who have spent their careers on these kinds of inquiries. What do you think you are likely to be able to add to their efforts in the short period of time that you have available for your thesis work? Not to mention the fact that the kind of information required comes largely from field observation data, which are notoriously resistant to standard statistical treatment. Usually, efforts of this kind end up being a bunch of anecdotes, 'just so' stories like how the leopard got its spots. Thesis committees don't like that, nor should they. Contrast this with our original plan. If you follow that, you will finish after two years in the field with a large amount of capture and specimen data, readily amenable to analysis. Those specimens in particular are like money in the bank—hard evidence of time well spent answering important questions."

Of course, he was right, and had I been raised in the maestro tradition typical of old-world-style graduate education, I probably would have acceded to my major professor's dictum. But I was not, and did not. Certainly, I respected his views and experience, but I was the one immersed in both the literature and data on a daily basis. In addition, there was the fact that, unlike most research grants, Welder awards were given to the student. True, a Welder Fellow was expected to work closely with his major advisor on the project, but the money and responsibility ultimately belonged to the awardee.

Warner was not the type to pitch a fit or throw his weight around in any case, and we were on good terms when he left, although he made it quite clear that he was entirely unconvinced and certain that I was wasting precious field time on a chimera.

Bonnie, too, was skeptical. By the second week in September, we were capturing over 150 migrants a day of twenty to thirty different species. Any person so intimately involved as she was with such a daily deluge of information could not help but wonder how to make sense of it. I had told her of Warner's misgivings, and she shared them. Her future depended on mine, and she too worried that the immense mass of capture data and observations would prove inscrutable. Naturally, I was worried as well. The broadening of the study and change in methods of data collection were both stimulating and frightening, especially since Warner had made plain that he thought I was headed in the wrong direction and wasting an exceptional opportunity (in not measuring and/or collecting as many birds as possible).

It is easy now, fifty years later, to suffuse in the golden haze of memory those early flounderings in search of the appropriate strategy and tactics to use in trying to understand a famously mysterious phenomenon. Fortunately, I have my field notes from that period to ground me. They are full of natterings and whinings to myself about the potential disaster awaiting my graduate career in failing to devise a methodology that could convince my committee that I had addressed a question of sufficient importance to warrant the granting of my degree. The path ahead was frighteningly obscure at the time, and by refusing to follow Dwain's guidance, I was stepping out onto it alone by my own choice.

I had hoped to perform a second experiment to test for territoriality in transients by using a tape recorder to play back songs and calls to see if I could stimulate a response in the form of approach and search for an intruder, as is often observed in response to playback by territorial males on the breeding ground. Accordingly, I had asked Warner to see if he could locate a Uher tape recorder (then the standard for professional field biologists) and directional microphone back in Minnesota. However, he had written me a couple of weeks after his departure to let me know that no such equipment was available, and besides, it was useless to be wasting my time on such frippery. Eventually (early October), Eric Bolen, Welder's new assistant director, found me a beautiful Nagra recorder with parabola and microphone—top-of-the-line equipment at the time—but I failed to get much response from the few transients that I was able to test.

In mid-October, I got a lesson in field biology logistical problems and the need to pay attention to local knowledge in learning to avoid them. On October 11, it started to rain at about one in the afternoon. I stopped by the office at about four, and Dr. Cottam's secretary, Rosemary Covington, asked me what I planned to do about my nets. I asked her what she meant, and she said heavy rain was predicted through the night and perhaps for most of tomorrow. "So what?" I said. She said that with rain like that, the Aransas River was likely to flood me out. I laughed and said, "No chance. It would have to rise thirty feet to cause me any problems." She just raised her eyebrows.

On the twelfth, rain continued throughout the day and night and into the thirteenth, by which time all of the nets were flooded, and two of them were completely underwater. Catch in the nets that I hadn't gotten to close when the water came over the banks included one garfish, three water snakes, five cottonmouths, and several sunfish and catfish. By October 14, the rain had stopped and the river was falling rapidly, but it took us more than a week to get all the nets cleaned and back in position.

Rosemary just smiled.

Common wintering birds began to arrive at Welder by mid-September, including Brown Thrasher, House Wren, Winter Wren, and Eastern Phoebe—earlier by a month than I had expected based on reports in the literature. We were busy handling the huge number of transients that were captured in our nets and conducting the little experiments with them described above. Nevertheless, the few observations I had the chance to make of the behavior of these wintering migrants were quite striking. Members of all four of these species were solitary, vocalized regularly (song and chip notes in phoebes and wrens, chirrs in the thrashers), and appeared to be defending territory.

As I have said, I had no time to work with them, but I did devise one little experiment to test for territorial response just as we were getting ready to depart for southern Veracruz. There was a House Wren living in the bushy plantings around the Welder Study (a large room with cubicles attached to the main administration building by a portico). I captured and banded the bird, releasing it back into its home bushes along the study wall just outside my cubicle window. I then brought a House Wren that I had captured in Hackberry Mott and tied it to a nail on a board, which I placed on top of one of the bushes outside the study—essentially, a literal "stakeout." It did not take long for the resident banded House Wren to discover the "intruder," which it attacked immediately and continuously for a minute or two until I ended the experiment by intervening and releasing the doubly abused captive. I filmed this event with an 8 mm camera, and although the quality was terrible (you could barely make out the two principals), I at least had documentation.

There were four striking aspects relevant to the wintering migrants arriving at Welder. First, as described above, they appeared to defend winter territories in stable, subtropical habitats—riparian forest and chaparral (thorn forest). Second, all four species studied are monomorphic: males and females are identical in appearance, and it was likely therefore that both sexes, and presumably different age groups, were defending individual territories. Third, the banding and recapture of several of the wrens and thrashers (phoebes usually foraged high above net height) showed that they were not wandering; rather, they remained near their original point of capture. Fourth, there were potential "ecological counterparts" indigenous at Welder for at least three of these wintering migrants: the Bewick's Wren, found in the same mesquite thorn forest habitat as the House Wren; the Carolina Wren in the same riparian habitat as the Winter Wren, and the Long-billed Thrasher for the Brown Thrasher (although the Long-billed Thrasher occurred more commonly in thorn forest, while the Brown Thrasher was usually in riparian forest). Recalling MacArthur's prediction, these native, resident potential ecological counterparts would be expected to deny access to stable resources in the principal habitats of the wintering area of the invading migrants, forcing them to wander throughout the nonbreeding period. Based on banding, recapture, and observation, the expected exclusion of these migrants from riparian and thorn forest habitats by their resident relatives did not happen.

As we packed up to leave for Veracruz, a disinterested examination of what we had learned during the first fall season at Welder would have to led one to conclude . . . not much. Although my thinking had been stimulated by the observations and questions discussed above, the various tests of those questions amounted to little more than ineffectual noodling, in no way decisive—an unfortunate attribute of a great deal of field biology, wherein the assumptions one makes in conducting a test often are unrecognized or invalid. I had little time to investigate these relationships further before our departure for Mexico, but they stimulated a great deal of conjecture on my part that was excellent preparation for work on wintering migrants in the Tuxtla rainforest.

PLAYA ESCONDIDA

Call me Ishmael.

—HERMAN MELVILLE, *MOBY DICK*

We left Welder on our long drive down to southern Mexico on the morning of October 29, 1973, accompanied by Bob Zink, a Minnesota undergraduate who had been helping us for a month or so. We stayed that night in Brownsville on the frontier and crossed into the border town of Matamoros the next morning. Reading back through my notes for that first entry into a new world for Bon and the kids, I am struck by my insouciance. Clearly, I was focused intently on each step forward along the road, attempting to avoid or at least mitigate disaster, without pausing to contemplate the momentous launching out our trip represented for them. Like notional migrants, they were wanderers, headed into the unknown. The kids were young, of course—Brigetta was four and Jay was two—so maybe the adventure was just another new part of growing up for them. Not so for Bonnie. She was twenty-seven years old, and although she had undergone some tough experiences as a single mom during our war years, they were not radically different from what many young American women of our generation had gone through. Now, however, she was headed literally and figuratively off the charts, traveling across a primitive landscape of burro tracks, log bridges, and poled ferries peopled by folks whose language she was only beginning to learn. Nothing in her previous life could have prepared her for this. She was a small-town girl who had married her high school sweetheart with the expectation of his eventual settling down as a physician somewhere in the vicinity of her natal community. Where she found the moxie to meet the challenges of managing a family in a third-world environment, I don't know. It had to be in her genes, I guess. All of her immediate family were immigrants—her dad a second-generation Swede whose parents had immigrated from Småland, her mom the daughter of immigrants to northwestern Pennsylvania. Bon's English grandmother's story seems especially relevant. When I knew her, she was about four feet ten inches tall and in her late seventies. She had been orphaned at the age of fourteen. She had gone to work then as a char lady, emigrating from her home on the Isle of Wight to Canada by herself at the age of nineteen, traveling in steerage. She married a Bavarian immigrant farmer called Hoffman, and they settled on a farm outside Warren, Pennsylvania, not far from Jamestown. She bore eight children while helping her husband run that farm—five boys and three girls (including Bonnie's mom, Margaret). If there was anyone tougher or more self-reliant than Victoria Eleanor Maude Gosden Hoffman, I don't know who they might be. Bonnie had that kind of fiber, and as we crossed into Mexico with our two small children on that autumn afternoon, she was going to need it.

Once leaving the outskirts of Matamoros, we left most evidence of the twentieth century behind as well. In fact, thatched huts and burro tracks were the chief evidence of humanity along the grandly named, sparsely traveled, and thinly settled Pan-American highway. There was no electricity, no running water, and of course, no phones. The road south from there crosses the vast, flat expanse of the coastal plain of northern Tamaulipas for the first hundred miles, then rises up low escarpments out of the flatlands and into a strange moonscape of low, rugged hills and mesas.

Figure 6-1. Tamaulipan mesas.

59

We stayed the night at a hotel in Ciudad Victoria, and the next morning, we walked to a nearby *restaurante*, a small establishment with three or four tables just off the main street, for breakfast. The food, *huevos rancheros, jugo de naranja, pan dulce, café con leche,* and fresh papaya with lime slices, was delicious. The chickens pecking around the floor at our feet, however, were a completely new experience for Bon and the kids.

After breakfast, we headed south. I wanted to show some of the country to Bon and Pug (my son's appellation for Bob) and so went on through to Mante and Valles, hoping to take the cutoff to Tanquian and from there to El Higo and Tempoal, then finally to Tuxpan for the night. All went well at first—gorgeous piedmont scenery with the peaks of the Sierra Madre beyond. When we arrived at the gravel road I thought was the turn-off to El Higo, I stopped to ask a seated *campesino* seemingly waiting for the bus (back in the 1970s, buses kept regular routes along nearly any road that was marginally passable) if this was the road to El Higo. He affirmed that it was. What he failed to mention was that the ferry across the Río Moctezuma had been swept away by the recent heavy rains so that while, yes, this was the road to El Higo, the closest we would be able to get to that town was about two hundred feet—the width of the river. This situation forced a fifty-mile backtrack and two-hundred-mile detour to Tamuin, thence at last to Tampico on the Veracruz-Tamaulipan border into which we finally straggled at six that evening.

We were not impressed with the city, although our long day in the saddle might have had something to do with it. My journal notes state, "Arrived at Tampico about 1800—noise, traffic, expensive, everything I hate about America right here in Mexico—even Colonel Sanders, the DQ, and Holiday Inn. Stayed at the San Antonio Motel on Avenida Hidalgo—mediocrity at its mediocrititist [*sic*]. Can't wait to head south."

The next morning, Día de los Muertos (November 1), we set out, taking the massive ferry in a dense fog of diesel fumes along with a bunch of huge trucks and buses across the Río Pánuco at about nine that morning. The road south from there to Tuxpan was paved but bad. I learned quickly on this route to pay serious attention to white rocks. At that time here and everywhere else in Mexico that I have traveled, a rock about the size of a coconut painted white and lain on the roadside was a standard signal for some issue ahead. The problem might be minor, like a pothole, but it could also mean complete blockage by landslide or treefall or overturned truck. Best to proceed with extreme caution.

After Tuxpan, the road was quite good to the oil town of Poza Rica (rich hole). From there, we headed to a stop at the ruins of El Tajín. Here we had a most amazing experience. I provide a little background below for you to have some understanding.

Figure 6-2. The ferry for crossing the Río Moctezuma at El Higo as it looked in February 1973 prior to its disappearance during the flooding of October of that same year.

Built around AD 800 by Totonacs, an indigenous people with their own unique language, El Tajín was the chief religious and commercial center of the region, home to an estimated fifteen thousand to twenty thousand people at its height. The metropolis was conquered and destroyed eight hundred years or so ago by the Aztecs, although an estimated eighty thousand Totonac people remain in the area today. One of the signature customs of the Totonac culture was the Danza de los Voladores (Dance of the Flyers). This ritual was developed to encourage the gods to bring rain during periods of extreme drought. It involved the placement of a pole one hundred feet in height, on the top of which was a revolving platform to which four hundred-foot ropes were attached. Five colorfully dressed "dancers" perform the ritual, four of whom climb the pole and attach a rope to their feet. The platform then begins to spin, and the ropes that have been wound around the pole play out, allowing each tethered dancer to approach the ground, gradually swinging farther and farther out away from the pole. The fifth man is perched quite precariously atop the pole. He dances and plays a small flute and drum while the others spin through the air.

As you can imagine, this performance likely was always a special event even a thousand years ago when the Totonac culture was at its height. Nevertheless, the Aztecs, who, like the Romans, were quick to appropriate attractive bits of culture from conquered peoples, evidently encouraged this ritual, as did the postconquistador Catholic Church, and it has persisted to the present day. Still, it is not a common occurrence. Yet as luck would have it, a dance was being performed at the precise moment of our arrival thanks to a generous contribution, I suppose, from a BBC film crew to the Totonac Cultural Preservation Society or some such entity. In any event, there it was. It was a warm day with clear blue sky above the dancers. We were the only people other than the three or four BBC folks in attendance. The flute and drum echoing through the noonday quiet and out across the pyramids as the dancers swung out farther and farther against the deep-blue vault above radiated a remarkable sense of timeless awe. I don't know about the gods, but it impressed the heck out of us.

Continuing south, we headed toward San Martínez de la Torre. About a quarter of the way there, just past the bridge over the Río Tecolutla, the paved road gave way to a gravel one—mostly giant gravel, three or four inches in size. We stopped after about ten minutes of creeping along through this mess to ask a *campesino* walking along the way how far it was to the town. He said about fifty kilometers. Well, there wasn't much we could do but keep going and hope that all bridges were intact and no recent landslides had blocked the way. It took about two hours to get back onto the paved Martínez de la Torre–Nautla road. Finally, we arrived in Nautla at about six in the evening and had supper.

At the restaurant, our towheaded children drew quite a crowd. It seems that touching the head of a blond-haired child was considered good luck, and several of our fellow diners wished to take advantage of the opportunity. The kids thought it was pretty weird but accepted the custom with decent grace, fortunately, because it happened a lot in public places over our next two years of Mexican visits.

After supper we had a long discussion regarding whether to stay in the sole small hotel in Nautla, eventually deciding in its favor—fruitlessly, as it turned out. Día de los Muertos is a national holiday, and all rooms were taken by families flocking to the coast for a seaside vacation. So we backtracked to San Rafael and grabbed the last two rooms available in that city at the Hotel Meuneir. *Room* turned out to be a bit of a euphemism for the attic closet that Bon and the kids and I occupied, as was *bed*—a low platform with some sort of shaggy material covering the wood that we shared with a distressingly large variety of active insect life. The grackle chorus serenading us from the *zócalo* throughout the night added to the singular atmospherics of our little dormitory. All four of us remember this stay quite clearly half a century later.

Figure 6-3. Market Street, San Andrés Tuxtla, November 1973. Obviously, the burro owner decided "No Parking" did not apply to him.

Despite our lack of repose, the spectacular sunny morning buoyed us wonderfully. The views along the coast road from Nautla to Villa Cardel are as beautiful as can be seen anywhere. We stopped for lunch on the *zócalo* in the historic port of Veracruz and then pushed on through Alvarado, across the bridge over the Papaloapan River, and into the ancient gateway town of the Tuxtlas, San Andrés Tuxtla, founded in 1532.

We pulled into Playa Escondida at about 3:30 that afternoon. As mentioned above, Día de los Muertos is a major holiday in Mexico, and just as vacationers had filled all the hotels along our route, so it was for the *cabañas* at Playa. As a result, Bonnie and the kids and I ended up sleeping on crates for a couple of nights in a storage shed shared with some obstreperous rats until the holiday crush cleared out.

Hotel Playa Escondida, as I have mentioned, was a business concept of Raul and Julieta Garcia. Originally from northern Mexico (Chihuahua), Raul was a bulky, balding, mustachioed man of sixty who habitually wore a worn khaki shirt and pants, a grizzled three-day growth of white beard, and an ironic half smile that seemed to say "I know life's not funny, but if you squint hard, you can see some humor in it." Julieta, a trim, dusky woman of forty or so, was originally from Mexico City. Unlike Raul, Julieta seemed

Figure 6-4. Our initial "living quarters" at Playa Escondida, subsequently converted to our banding and specimen prep lab.

to find little humor in life. The hotel part of the operation was her responsibility, and she ran a tight ship.

They owned a small cattle ranch that happened to have a cliff overlooking two spectacular beaches: a broad, curving white sand beach to the south, and a smaller black sand cove (the eponymous "hidden" beach) to the north.

Grasping the entrepreneurial possibilities of the site, they built an open-air, patio-like restaurant with their living quarters above and a large tree growing in the middle up through the ceiling. The guest housing consisted of two (eventually, three) concrete, single-story buildings with flat roofs of corrugated iron. Each structure housed four individual units consisting of a bedroom and a small attached bathroom with

Figure 6-5. The hidden beach.

sink, toilet, and shower. Water came from a covered cistern located on a hill a hundred yards or so west of the rooms, which the cows fell into only occasionally. No hot water, of course.

Raul and Julieta were assisted in running the ranch, restaurant, and hotel business by a family of Indians of the tribe indigenous to the Tuxtlas, the Popoluca—namely, Angel and Francesca Toto and their five kids: Gloria, 11; Maria, 8; Virginia, 6; Domingo, 4; and Anna, 2.

Once we were out of the rat house and settled in our actual guest room, my family adjusted quickly to their new surroundings. Bonnie assumed duties as data recorder whenever Mario, Dick, or I brought birds to be banded back to the lab, and when she wasn't doing that, she helped out Julieta and Francesca in the kitchen. Also, once word got out that she was a nurse (*enfermera*), anyone ill or injured from our little group or the nearby fishing village of Jicacal made their way to her so that she essentially operated an ad hoc clinic. During our stay, she delivered Francesca's sixth child, Marguerita, named in honor of Bonnie's mother.

The kids also adjusted quickly to their novel environment, spending their days playing with Maria, Virginia, and Domingo, from whom they learned what they should and

Figure 6-6. Virginia Toto holding her sister Marguerita (Francesca Toto's new baby, named in honor of Bonnie's mother) and my son Jay.

should not do, such as how to harvest and prepare (by charring) the nuts from chocho palms (a spiny understory tree), how to recognize and avoid the army ant swarms that occasionally invaded the hotel units for a few hours, or how to compete for the boiled chicken feet discarded after the daily preparation of broth for the evening meal. In fact, they developed a certain insouciance resulting from their obviously superior knowledge of the daily workings of the place, which revealed itself on those occasions when I demonstrated a pathetic lack.

As, for instance, when I was kicked by the bellwether cow. A herd of about ten cows came up from the pasture to the back of the patio every morning for milking. I ignored them completely, walking back and forth to the lab right though the herd until one day, the lead cow gave me a hard kick in the hip. The kids just looked at me like I was some kind of an idiot. Everyone knew that all the cows were harmless *except* her, who was spiteful and should be avoided.

Similarly, they were not surprised when things went bad between Raul and his "pet" coatimundi. He had been given this animal by a hunter who had killed its mother. Raul kept it tied up to a tree next to the patio where he went betimes afternoons after *comida* to commune with it and feed it scraps. The kids never went near the thing, knowing full well that it was vicious. One evening when we were heading over to the patio for dinner, I noticed that the coati was absent.

"Where's the coati?" I asked.

"Oh, well, it bit Raul this afternoon, and he beat it to death with a stick," they explained.

The kids also developed a working Spanish vocabulary that was much larger than mine. When I wanted to know a domestic word—like *socks*, for example—I simply asked them rather than looking it up. "Calcetines!" they chorused. Unfortunately, their psittacine capabilities were not restricted to a household lexicon. They picked up some unsavory, if sonorous, phrases from the construction workers who were building Raul's new *cabañas*, which they chanted gleefully until discouraged. Mario was too embarrassed to translate, so I will leave them to the reader's prurient imagination.

I had a bit more difficulty fitting into Playa's professional protocol. My colleagues on the project, Dick and Mario, had arrived with Dwain in the Tuxtlas back in July to set up the tropical portion of our project, establishing their housing arrangements with Julieta and a site at the UNAM field station. Although the original idea had been for the two master's student projects, undertaken by Mario and Dick, to be coordinated with

Figure 6-7. My children, Brigetta and Jay, with a baby howler monkey. Our *cabaña* at Playa Escondida is shown in the background.

mine so that their work could be done while assisting me, things did not work out that way. Even though all three projects involved mist-netting birds, the specific objectives for each were quite different.

Mario's project was to focus on the question of whether migratory birds of different breeding populations followed different routes, had different timing of migration, or used different wintering areas. This idea had been a part of my original proposal, but I had come to realize that I could not examine this question, which would require the exhaustive collection and preservation of specimens, and still look at the questions posed above regarding MacArthur's predictions for bird behavior during migration and on the wintering grounds. I needed live, banded birds to address those.

As for Oehlenschlager, his work was to be a broad examination of the ornithogeography of the Tuxtlas, a study in which migrants would constitute only a minor part, less than 20 percent of the bird community of the region. In addition, their studies were to be conducted mostly at the UNAM Tuxtla field station, supplemented by ad hoc collecting by shotgun in neighboring fields, pasture, and second-growth brush. Warner planned for them to run mist nets there from August 1973 to May 1974 and from August 1974 to May 1975. No work for either of them would be done in Texas.

Thus rats were the least of my problems so far as our arrival at our Tuxtla study sites was concerned. Warner had established sampling methods that focused almost entirely on inventorying the avifauna of tropical rainforest, mainly by collecting and preparing specimens mist-netted at the UNAM biology station. No main catalog recording each specimen captured had been kept. Instead, specimen data were written on the museum tags accompanying each specimen; banding records were kept separately. These procedures would provide few data pertinent to the questions that I considered important, so I sat down with Mario and Dick to change them.

First, I said, there had to be a single catalog where data for every specimen captured, whether collected or banded by anyone working on the study, were to be entered. Second, I wanted all migrants captured, except those critical for Mario's subspecies work, to be banded and released. In fact, most migrants that showed subspecific variation, like the Northern Waterthrush and Gray Catbird, were more common in habitats other than *selva* anyway; only the Swainson's Thrush was found regularly in forest. Accordingly, the best method to address Mario's question was to mist-net or shoot individuals of those species of migrants wherever they occurred and not waste time netting in forest. Third, I explained that I was not going to help with their netting at the station. The questions that I was interested in were best answered by capturing, banding, and releasing migrants and studying their behavior in various tropical habitats, including pastures and overgrown fields as well as *selva*. This work could be done in the immediate environs of the *cabañas*, obviating the necessity for the thirty-minute commute between our lodgings and the station six times a day.

Dick was furious and adamant that Dwain's protocol be followed. Although ostensibly a student whose thesis question arguably would have been best served by sampling major habitats in addition to *selva*, he was basically Warner's technician. He had been given his instructions on how the project was to be run by Dwain, and that was that. He also agreed with Dwain that fieldwork in rural Mexico was no place for a wife and children, warning Bonnie and the kids of the dangers of poisonous snakes, diseases, and bandits. In fact, his welcoming story for them on our arrival was that thirty-nine fer-de-lances (a poisonous tropical relative of the rattlesnake without the rattle) had been killed on the compound. We later found out from Mario that this story was true, sort of. Julieta had a woman who helped Francesca, Angel's wife, with cooking and the laundry. She and her husband, Domingo, a subsistence farmer, lived in a shack with their two young daughters about a hundred yards up the road from the hotel. Domingo was a strapping, good-looking guy who spent most evenings drinking and brawling (two missing front teeth) at the nearby fishing "village" of Jicacal (in truth, more like a seasonal camp occupied chiefly by vagrants from Veracruz city). It seems that he had been reeling home late one night and come across a large fer-de-lance in the road, which he quickly dispatched with his machete. The next morning, he brought it to Dick, who opened it up and found that it was a female with thirty-eight young inside her—hence the "thirty-nine fer-de-lances."

Had we been operating under the maestro system, as discussed above, Dick, as Warner's on-site representative, would have had the last word. But we weren't, or at least I wasn't. So far as I was concerned, it was my project. Besides, Warner was two thousand miles away. Consequently, an uncomfortable agreement establishing the team's long-term working relationship was reached. A main catalog would be kept into which all captures would be entered regardless of what specific site they were from (Bonnie ended up being the chief recorder most of the time for all data gatherers during our tenure at Playa). Mario and Dick would continue commuting to the station to run their net lines, but most migrants captured would be banded unless Mario specifically requested that they be kept as museum specimens for his subspecies work. My study sites would be located in *selva*, pasture, and fields in the vicinity of the *cabañas*. I seldom went to the station.

Dick never warmed to any of this arrangement, nor did Zink, who considered himself Dick's assistant, although he helped me out a great deal in Texas and during the initial stages of setting up my net lines at Playa. Mario was caught in the middle. He was Dick's roommate, and they lived and worked together for most of the two-year period from August 1973 to May 1975. But Bonnie and I considered him, and his family after his marriage to Isabel in 1975, to be among our closest friends then and over the years.

Regardless of the logistical and personal issues involved, Playa turned out to be an ideal location to address the questions we had begun to attend to in Texas, although it took me some time to work out how best to do so. Once I got the family settled in our *cabaña*, I explored Raul's property in the immediate vicinity of the "hotel," guided by his manager, Angel, finding about ten acres of relatively undisturbed rainforest, two acres of rainforest in which the understory was pretty heavily grazed by cattle, and three or four acres of mixed pasture and low, scrubby second growth. On completion of the tour, I set about establishing a grid of mist nets that would sample the birds—initially, mostly in the rainforest. Because of the high canopy in the forest (well over 100 feet, with emergents to 150 feet), I thought that I should at least try to get the nets at various levels above the ground. This decision, while not a complete fiasco, ended up costing me a lot of field time that might have been best spent on other things. Experimenting with various methods, I eventually settled on a rope-and-pulley system to set these "sky" nets, as I euphemistically termed them. I tried different ways of getting the nets as high up as possible, but the only reliable method turned out to be climbing. I had no climbing irons or other equipment of that type and ended up using vines or nailing two-by-four-inch planks into tree trunks to serve as rungs in a ladder. The vines turned out to have drawbacks. First, no matter what their size and contrary to Tarzan's experience, they were entirely undependable. I recall one instance where after ascending forty feet by vine to set the pulley and descending back to the ground, the vine I had used, which was a couple of inches thick, simply fell out of the tree, landing at my feet! I attributed this event to a harmless warning by a local forest spirit—called *chaneques* in the Tuxtlas. Such beliefs are scoffed at, of course, by colleagues, but after forty years of wilderness work, I've found you can't afford *not* to pay attention to them. The second problem with vines was that some contained caustic sap that burned like fire if you got it on your skin, which I always seemed to do. A third problem was that I found it difficult to climb in boots and pants, so I usually stripped down to skivvies for the ascent, much to Zink's amusement. Of course, greenhorn skin is not used to such abuse, so I always got nicks and scrapes in various places, one of which was a brush burn on the top of my foot suffered on November 17. This wound ended up getting infected, and as often happens in the humid tropics, it took a long time to heal. It was not until November 30 that I could put on a boot.

No matter what method I used, the highest I could get the nets was fifty feet or so. I could increase their effective height by locating the nets in trees on a ridge, but there were only a couple of places on Raul's property where that would work. The biggest problem with sky nets, though, is that they are far less efficient in capturing birds than ground nets but far more effective in catching sticks and leaves. Nevertheless, I persisted, with Zink's help, in establishing twelve of these "rigs," so-called because simply calling them nets belied the complex system of ropes and pulleys used at each site to raise six or seven nets, forty feet in length, from a cradle near the ground up into lower parts of the canopy forty or fifty feet above ground—a virtual wall of netting. One reason these rigs were less efficient than ground nets was that the birds could see them more easily, especially if there was the slightest breeze. Also, it is difficult to get the tension just right between the support ropes, which is critical for capturing the bird once it has hit the net. As the weeks passed, I gradually added more and more ground nets in the vicinity of each sky net so that I could be sure to capture and band as many migrants as possible that were winter residents on the property, over half of which spent most of their time in the understory, where they were captured much more efficiently by ground nets. By the third week in December, I had pretty well covered Raul's ten acres of *selva* with a grid composed of one net or rig set every hundred feet.

Each migrant captured was given a unique color-band sequence for individual identification in the field prior to release, enabling me to document by observation what birds lived where. Based on this information, the migrants found in *selva* seemed to fall into two broad categories in terms of their behavior and distribution: solitary birds that could be seen usually in the same piece of forest all day, every day, and birds that associated with the mixed-species flocks of resident birds that roamed at midcanopy level across several acres of forest. The recapture and observation of color-banded birds documented that many individual migrants of several species spent the entire wintering period on or near the same small piece of forest.

Individuals of nearly all of these species were heard to vocalize regularly—that is, every hour or so—throughout the day, with peaks in midmorning and early evening. Most used chips or chirrs for this purpose, although song also was used occasionally by the Black-and-white Warbler, White-eyed Vireo, and Yellow-bellied Flycatcher. Mario was able to document for the vireo (by shooting in flagrante) that both males and females sang. Using the Nagra recorder with the parabolic microphone, I recorded examples of most vocalizations given by migrants, which I used in playback experiments to see if I could get a response (an approach to within a few feet of the recorder), like what one might expect from a male bird on its breeding territory. There was quite a range of responses to playback, depending on species. Hooded Warblers responded rapidly by approaching and searching for the source of the sound more than half of the time; other species, like the Wood Thrush and Kentucky Warbler, responded some of the time; while still others did not respond at all.

The site fidelity and vocalization observed in most species of migrants wintering in Raul's rainforest were indicative of territoriality. Even those birds like the Magnolia Warbler and American Redstart that were seen mostly in mixed-species flocks usually were represented by a single member of their species in these flocks, appearing to defend against membership in the flock by conspecifics (members of the same species). In addition, on a few occasions I observed obvious territorial encounters between free-flying individuals of a few migrant species. Nevertheless, such sightings amounted to little more than suggestive anecdotes, and I was eager to see if I could produce clear-cut territorial responses on demand. To this end, I developed two methods for introducing a potential "intruder" onto a suspected territory. The first involved the same technique I used for staking out the House Wren back at Welder—namely, tying the bird by thread to a nail on a board—while the second involved the construction of a small cage a cubic foot in size built of wire and mist-netting with a wooden floor. At first I would just walk out in the woods and choose a likely place near where I had seen a vocalizing member of the target species. However, except in the case of Hooded Warblers, I found that my presence inhibited responses for individuals of most species. Therefore, I built a portable blind made of burlap so that my presence would be less obtrusive. From late December until late February, I spent about 150 hours in that blind with my camera ready.

Thank goodness for Hooded Warblers. If you stake out a Hooded Warbler, caged or tied to a plank, on the territory of another hooded, the territory owner will respond immediately. It will approach, give stylized displays that are the same for every territorial individual tested, and ultimately, attack when the intruder doesn't leave. Furthermore, the owner doesn't tire. As long as you leave the "intruder" in place, it will continue to vocalize, display, and attack. Fifty minutes was the longest time for which I left a caged bird in place. The owner was still attacking when I took pity on both of them and broke it up.

Other species were much pickier. Something about the experimental setup seemed to deter them. Nevertheless, I was able to observe and photograph territorial responses, including attack and stylized displays in individuals of several migrant species in defense of their rainforest territories.

Migrants wintering in the rainforest understory in the Tuxtlas show different distributional patterns in different places, presumably depending on the kind of microhabitats available. For instance, Louisiana Waterthrushes are found only along arroyos and Gray Catbirds only in thickets. In Raul's *selva* there were really only two species of migrants that had fairly uniform distribution with territories throughout his forest: the Wood Thrush and the Hooded Warbler. The Wood Thrush is furtive and does not always respond well to caged intruders or playback, but the Hooded Warbler proved to be quite dependable. If you walked out to where you thought Yellow Left (his color-band sequence) was supposed to be, you could nearly always find him, or, if not, you could bring him in with playback. Also, hoodeds are not furtive. They forage three to six feet or so above the ground by sallying for flying insects and usually don't change their behavior much when approached by an observer.

These characteristics made me wonder if I might not be able to map the boundaries of their territories. We did not have GPS back then, so all of the location points had to be marked down on a hand-drawn map made with a compass, a hundred-foot measuring tape, and pink flagging. Using these items, I set out to map the territory boundaries for ten Hooded Warblers I knew from banding and observational data to be territory owners in Raul's *selva*. I used three methods to obtain points: (1) recaptures and resightings, (2) responses to playback and stakeouts, and (3) use of the "territory flush" technique developed by John Wiens to determine boundaries for grassland bird breeding territories. Most of the data, however, came from the flush technique, the others being far too time-consuming. This procedure works for Hooded Warblers because of their visibility, uniform distribution, and relative tameness. It involves two "beaters," usually Zink and me, slowly walking through the understory, "pushing" the territory owner toward his boundary. When he gets there, or a little ways beyond it, he will fly back over or around the beaters, often after being attacked by his neighbor. The point from which he flew was marked with tape and on the map, and the process was repeated until the boundaries for all ten birds were mapped.

To my surprise, Warner arrived at about five in the afternoon on New Year's Day 1974. He stayed for a month, running a few mist nets that he had set up near the lower *cabañas*, but spending most of his time drinking coffee on the patio and talking with whomever was back from the field at the moment. It was great to have him there. He remained deeply skeptical about my behaviorally centered approach to migrant study, but that was not a bad thing. In fact, it is good to have an interested, knowledgeable inquisitor questioning your methods while you are pursuing them, and on January 26,

I was able to take him out into the nearby *selva* and show him the single-minded ear-nestness with which Hooded Warbler Red Left attacked Yellow Left in his mist net cage. The interaction continued for nearly an hour, with Dwain and me in the audience until I stepped in and released poor old Yellow Left from his misery. Red Left remained in doubt, however, and continued attacking the empty cage for an hour after Yellow Left had departed. When I had described these kinds of interactions in our previous con-versations, Dwain had always pointed out that all of my accounts could be explained as simple interest in, and inspection of, the odd "caged-intruder" setup. Now he got it. Migrants were *not* wanderers and were *not* excluded from stable tropical habitats—a *Eureka!* moment of the first order.

Dr. Cottam and Eric Bolen arrived at Playa at about noon on January 17. As was the case with Dwain, I had no idea that they had planned to visit but was delighted to see them. They went out with me to observe my netting sites near Playa and also with Dwain, Dick, and Mario to see the net lanes at the station. They stayed only a few days, spent with one or the other of us at our study sites or exploring and taking pictures on their own. They both had terrible cases of chiggers, which, as Dr. Cottam commented, "seemed to have no respect for a man's pride or passion." He often looked tired but was in excellent spirits and got out into the field every day with Eric. I remember standing on a hillside with him overlooking a spectacular view of pasture and forest with the cliffs and sea beyond. He paused, leaning on a walking stick, and said such a scene made him glad to be alive and that he wouldn't mind if he died tomorrow. As it happened he was dead from prostate cancer within three months.

I presumed that the reason Eric and Dr. Cottam had visited us was because they were intrigued by the exciting work about which Warner and I had told them. Certainly, that is the line they gave to me. Later, back up at Welder, I heard a rumor that one of my graduate student colleagues had commented to Eric that it seemed pretty strange for Dr. Cottam to trust a notoriously dissolute university professor and his three students of notably undistinguished academic background to do anything worthwhile down in Mexico with all that grant money. The person relating this said that this discussion wor-ried Eric enough that he went to Dr. Cottam and expressed the thought that it might be worth an official inspection of our supposed Mexican study area to make certain that something of value was being done with the considerable funds invested by Welder.

During the last week in February, I wrapped up my work at Playa and prepared to pack up the family and head back to Welder to begin our spring field season in Texas at Hackberry Mott. What we had discovered about migrant biology in the Tuxtla

Figure 6-8. Dr. Cottam helping me remove a Hooded Warbler from a mist net at our Playa study site, January 1974.

rainforest confirmed (for me, at least) what I had begun to suspect during our fall work in Texas—namely, that the northern home paradigm for understanding bird migration was wrong. Migrants were not wandering interlopers in the southern regions in winter. Members of both sexes of at least twenty species of migrants arrived at their tropical homes on our study sites in the fall and remained on the same small piece of ground for several months until their departure for their temperate breeding areas in spring. Furthermore, these pieces of ground were sufficiently important to them that they spent whatever time was necessary in the advertisement and defense of a territory, preventing other members of their species from using it as well. In addition, there was no evidence that bird species resident in South Texas or southern Veracruz had any effect whatsoever on migrants wintering in their communities, let alone serving as "ecological counter-parts" to their migrant relatives and excluding them from tropical habitats. Somehow, wintering migrants behaved as though they *belonged* to their winter communities.

But how could this be so?

SPRING AND EARLY SUMMER 1974

The difference between warblers
and no warblers is very slight.

—Frances Lee Jacques, quoted in Florence Page
Jacques, *Artist of the Wilderness World*

We left Playa Escondida at about noon on February 23, heading for Mexico City for administrative visits with Bernardo Villa and Dr. Gomez-Pompa at the Instituto de Biología and Ticul Alvarez (former student of Bernardo Villa) at the permit office. I had brought with me several hundred bird skins collected and prepared during our first year of fieldwork in the Tuxtlas and needed to get an export permit for these from Ticul as well as renew our permits for the coming year. He asked that we leave some portion of our specimens at the Instituto's bird collection, so I headed over to inspect their facilities.

I had expected to find Allan Phillips there and to be shown the fruits of his decades of fieldwork in the form of a large and beautifully curated collection. The reality was sadly different. Instead of Allan, I met with curatorial assistants Isabel Castillo (soon to be Mario's wife) and Lourdes Trejo. They explained that Allan had had a falling out with the institute's higher-ups and decamped for Monterrey, taking most of his personal collection of bird skins with him. What remained was about a thousand badly curated specimens consisting mostly of the leavings of earlier gringo investigators, such as W. W. Brown, Alexander Wetmore, Bob Dickerman (Warner's former student), and of course, Allan Phillips. There was no main catalog, and individual specimen labels often contained no data other than the collector's personal catalog number. Preservatives for the specimens were nonexistent. There was no money, Isabel explained—a horrifying situation unlikely to encourage collaborators to make large specimen donations.

We stayed a couple of more days in the city, sight-seeing and paying social calls, before heading on to Monterrey in hopes of finding Allan in his new home. In this we were successful, pulling up in front of his house on a dusty road in the industrial Monterrey suburb of San Nicolás de los Garza at about 10:00 a.m. on March 1. Allan was there along with his wife, Juana, and they welcomed us warmly. The living areas were neat enough, but Allan's large office looked as though it had been the recent subject of a thorough search by the KGB, with books, papers, specimens, and collecting gear scattered seemingly willy-nilly. Noticing my open mouth and staring eyes, Allan reassured me that while he was still in the process of getting organized after the move, he knew where pretty much anything of importance could be found.

Taking him at his word, I brought in specimens taken at our Tuxtlas sites of seven migrant species known to show recognizable subspecific variation in plumage coloration by breeding area to see what he could make of them. I should mention that no other person in the world at that time could have made any guess regarding subspecific identification of these birds without a large reference collection at hand. Allan was happy to make tentative IDs for all of them.

Once he had completed this task, I asked him what he thought about the probability of different wintering areas for different subspecies, to which he responded that it was nearly always the case based on his own experience. This view, along with his IDs, made the trip to visit him worthwhile indeed, since the concepts and methods discussed formed the heart of Mario's thesis work. I also told him about my findings regarding migrant territoriality in tropical rainforest, which, to my astonishment, he said was not surprising to him: Wetmore had reported it for American Redstarts and Lincoln's Sparrows in his publication on the Tuxtlas avifauna back in 1943. I did not press him further, but his comments in the Smithsonian symposium on new-world migrants, edited by Helmut and Jimmie Buechner and published in 1970, indicated that he agreed, or at least did not disagree, with MacArthur and others that migrants were nomads in the winter season.

After lunch as we prepared to head on up the road, Allan asked if I would mind taking a few of his birds with us, since we already were taking a bunch across the border. He explained that after his falling out with the Instituto, he had decided to donate his collection to the Delaware Museum, where his good friend and colleague John Hubbard was curator. Even though the "few birds" turned out to be four large boxes containing over a thousand specimens, I said, "Sure. No problem."

But there *was* a problem.

Our previous border crossings had been at the busy Brownsville-Matamoros interchange, but the quickest route from Allan's Monterrey home to Welder was via the

Reynosa-Hidalgo interchange, a much sleepier place, at least until our arrival. Negotiating the border heading *out* of a country is not difficult under most circumstances. Usually, to get out of Mexico, all that was necessary back in 1974 was a stop at the customs office (*aduana*), where you drop off your visas and go on your way. However, as I got out, a customs official came walking around the back of the truck and asked me, "¿Qué tiene en las cajas?"

"Pieles," I said.

"¿Pieles?" he said rather loudly.

Skins, evidently, was not the answer he had expected to his question of what was in the boxes, and things went rapidly downhill from there. He told me to get those boxes out right now and show him exactly what we had, which I did. By the time I had all of my boxes and all of Allan's out and opened, a small army of customs and immigration officials had gathered, including, unfortunately, the *comandante*, who was grumpy from being summoned from his postprandial siesta. A difficult conversation, all in Spanish, ensued. Admittedly, my language skills had progressed beyond the *cerveza*, *biólogo*, and *baño* stage, where they had been when I had my encounter with the good citizens of Balzapote, thanks largely to some rather intense discussions with Señora Julieta regarding our room rent and meal costs. Nevertheless, they were severely stretched. The whole issue of specimen export was discussed and debated at length. No one was up on the latest regs, so the *comandante* ordered the customs bible to be fetched, which turned out to be a thick, dusty tome with crinkled yellowed pages, where he found on page 552 under chapter 11, section 7, that the export of specimens required permission from the rector of the university or his or her designated representative.

The following represents a rough translation of the ensuing conversation:

COMANDANTE: You must have permission from the rector of the university to export specimens.

ME: Fine, no problem. Call him.

COMANDANTE (WITH A LOOK OF HORROR): Call him? Call the rector?

ME: Yes. The rector. I have done this before without problems. Call him.

The *comandante* stared at me for a moment. Then he stared at his subordinates. Then he motioned to them for a sidebar. Then he turned back to me, continuing our discussion:

COMANDANTE: You have done this before?

ME: *Yes.*

COMANDANTE: Where?

ME: Matamoros.

COMANDANTE: Why don't you go on to Matamoros and cross there?

ME: Good idea.

With that, the *comandante* and all of his attendants left but one. The sense of my discussion with this official goes as follows:

OFFICIAL: You know, your children are not mentioned on your visa.

ME: Yes. I'm sorry. I had no idea that they had to be.

OFFICIAL: Of course they have to be. They are here illegally.

ME: Would fifty pesos solve the problem?

OFFICIAL: This is not a joking matter. I could have you and your wife arrested for their illegal entry and for offering me a bribe.

ME: Well, then, I guess you would be responsible for the children.

OFFICIAL (NOW PLAINLY FURIOUS): You should have been detained both for the illegal specimen export and the visa problem.

I had no response to that. He was wrong, though, and the *comandante* knew it. His (the *comandante's*) thinking probably went like this: *Detaining an American citizen, his wife, and two little children for a bunch of dead birds and a visa oversight has the potential to cause far more problems than whatever satisfaction this American's discomfort might produce. The right thing to do is to impound the birds. But what if this guy really does know the rector? No. Getting him out of here is the best solution for everyone.* Or almost everyone.

OFFICIAL (IN RIGHTEOUS INDIGNATION): Get out, and don't come back!

We left, heading south and east along the roundabout way to Matamoros, which added a few hours to our trip, but at least we weren't "detained." And I have not been back to Reynosa since.

When we got to Matamoros, I told the children to lie down on the floor and be quiet while I stopped at the *aduana* to turn in our visas. If they were discovered, I explained, they would have to stay in Mexico on their own. They got down on the floor and were quiet as four-eyed possums (a small Neotropical marsupial, normally very quiet). All went well. I turned in the visas, and we crossed the Rio Grande into the United States. No problems on the US side. We had all the papers necessary for importing the specimens, and they didn't care about visas or kids, so we were on our way in about five minutes. *Whew!*

We arrived back at Welder at about two in the afternoon of March 2. Both Dr. Cottam and his assistant director, Mr. Glazener, were in the hospital: Dr. Cottam with terminal prostate cancer, from which he passed on March 31, and Mr. Glazener with a heart attack suffered during a controlled burn on one of the Welder pastures. Eric Bolen was running things day to day. Doug Mock, a colleague of mine from the Bell Museum, had taken up his fellowship at Welder to study heron nesting behavior and was living with his wife, Karilyn, in the dorm.

Gene Blacklock, curator of the Welder Museum, was good friends with Frank Oatman, a young professor at Goddard College. Gene had agreed with his friend to work with a Goddard student, Chris Barkan, during the spring semester to teach him about bird identification and curation techniques. The course of instruction was not going that well when we arrived back at Welder after our Mexican sojourn. Gene was certainly an open-minded Texan. Nevertheless, he found Chris's behavior and philosophy of

Figure 7-1. Gene Blacklock, April 1974.

life completely foreign. Goddard was one of only two colleges that I knew of to offer a student-designed curriculum in 1974; Evergreen was the other. Both were avant la lettre for their time.

Chris was a bright, normal, twenty-year-old Connecticut Yankee—into Captain Beefheart, Pink Floyd, free love, and so on—but his ethic was not what Gene considered to be proper. In his first evaluation for Goddard of Chris's performance, Gene wrote, "He is a constant source of irritation." Chris showed this to Bonnie, who thought it was so hysterically funny, she had it made into a T-shirt for Chris that stated "I am a constant source of irritation." Sadly, Chris did not think this humorous, so Bonnie kept it and wore it proudly for many years until it finally fell apart. In any event, Chris and Gene were not well suited to each other, so Gene suggested that Chris might be able to work with us, a solution suitable seemingly to all parties.

This offer proved to have outstanding long-term benefits. Chris turned out to be not only a hard worker and quick learner but a wonderful friend and colleague for

Figure 7-2. Chris Barkan helping me set up a mist net in Hackberry Mott, April 1974.

both Bonnie and me up to the present day. He and his wife, Libby, visited us this past summer (2022).

We were lucky to have him during our first spring season at Welder, not only because of his talents, but also because in addition to conducting my fieldwork, I had to prepare for my written doctoral preliminary examination ("prelims"), which was scheduled for sometime in May. As a result, Bonnie and Chris, with occasional help from Karilyn Mock, did most of the mist-netting, cataloging, and banding in that spring season.

When I was not studying or mist-netting, I was trying to develop and implement tests of territoriality in transients, involving various combinations of playback and caged-intruder experiments similar to those conducted in the Tuxtlas. The results were intriguing but nothing that could be subjected to statistical analysis. Regular vocalization, site fidelity, and occasional direct interactions observed in the field convinced me that some individuals of at least thirteen species occurring at Welder as passage migrants defended small feeding territories for a few days during their stopover. The problem with testing this response was that apparent territory holders (vocalizing, site faithful) often ignored conspecific intruders, whether free flying or caged. If you think about it, this makes sense. After all, most of the birds in Hackberry Mott are there for a single day and may not represent competitors for territory at all. Probably these birds are able to convey this message to territory owners through their body language and lack of vocalization. This is perfectly logical but not helpful in terms of building a case for territoriality at stopover sites for the rebuilding of fat reserves.

My doctoral prelim test arrived in the mail on May 9, whereupon I shut myself up in the Welder boardroom for the next four days writing responses to my interrogators, which Bonnie typed up. Once the test was shipped off, I was free to work with Bon mist-netting, but there was not much left of the season. We pulled the nets on May 24 and headed north for Minnesota on the twenty-fifth.

Most PhD programs at universities in the United States require the passage of tests commonly called the preliminary oral and written examinations for the doctoral degree. However, the term *preliminary* is somewhat misleading in this context, as it implies that these exams come early in the aspirant's graduate career. At the University of Minnesota, this was not true, at least when I was in attendance. The early exams were called "comprehensives." These consisted of two or three days of tests covering all major topics within the aspirant's field and were taken shortly after acceptance into the program. Their purpose was mainly to determine the principal weaknesses in the student's preparation so that a coursework program could be designed to repair those weaknesses. The

preliminary written and oral exams came later, usually after the completion of coursework but before one begins thesis work. Their main purpose is to weed out those deemed
not qualified for a doctorate. The written exam comes first, and if you pass that, then
you stand for the oral prelim, which, in most universities in my experience, is the most
important test you will ever take. If you pass the oral prelim, then it is assumed that you
will eventually get your doctorate. All you need to do then is perform the research and
write it up in a form that is acceptable to your committee. There is a final oral exam taken
after your thesis has been accepted, but this generally is valedictory. It is when you pass
the oral prelim that you can call yourself a candidate and use the important term ABD
(all but dissertation) in your job applications. Your committee teaches you the secret
handshake, breaks out the champagne, and welcomes you into the sacred precincts and
higher mysteries.

Due to various problems and issues having to do with my own peculiar circumstances as well as Dwain's, my prelims worked out quite differently. As I described in
chapter 1, I enrolled in the Bell Museum's graduate program during the summer of 1968,
taking my first classes in the fall quarter of that year. Like my fellow graduate students, I took the comprehensives early in the quarter. These showed weaknesses in statistics, biochemistry, and systems ecology, so Dwain and I set out a plan of coursework
covering those issues as well as other areas in which I was particularly interested over
the next couple of years. I took courses in statistics, biochemistry, and plant ecology that
quarter, but then I was drafted. When I returned three years later during the winter quarter of 1972, I was told by Dr. Tordoff, as I have related, that I would not be accepted into
the museum's PhD program, so I set out to do a coursework (nonthesis) master's, which
I could complete by the end of the year. But then I was awarded a major grant in the fall
of 1972 that would pay not only for my own PhD work but for master's degrees for my
two assistants. The grant made my nonacceptance into the PhD program moot. How
could they not let me proceed? I had completed my master's degree and initiated exploratory efforts for field sites in Texas and Mexico before I even had a committee, and I had
almost a full year of fieldwork before taking my written prelims in the spring of 1974.
In fact, as noted above, the test was sent to me while I was still in the field at Welder in
Texas. Nevertheless, my committee was pretty happy with the results, so I scheduled my
oral prelims for my return to the Twin Cities in June.

As it turned out, two of the five members of the committee asked to be replaced
before the oral prelim due to time pressures, so I now had two faculty who knew nothing
about me except my transcripts. The exam took place early on a Friday afternoon in a

classroom off from the museum's Touch and See Room—a large, open, windowless, and dimly lit space where teachers brought grade school kids for hands-on learning about natural history.

I have mulled over those two hours in my mind many times in the years since and still cannot recall much of note that occurred during the exam itself. The only question that I knew I had flubbed was from the biochemist on the committee. On the written prelim, I had been asked to identify and describe the accomplishments of seven prominent biochemists. I was unable to identify two of these guys, and this committee member asked me if I had looked them up and could now answer properly. I had looked them up in two recent biochemistry texts but had failed to find them, so I had to admit that I could not. I couldn't see how this would be a major problem, but as the time dragged on for me, sitting in the semidarkness outside the exam room with little brats screaming around me, I began to become seriously worried. The normal time for a decision is fifteen minutes or so, and as it closed in on half an hour, I began to suspect that I was in trouble.

Eventually my fears were confirmed when Warner came out alone, looking like death. "Well, I know what you are capable of, John," he said, "but the committee was unconvinced. They were unhappy with your failure to identify the biochemists, and Dr. C— (the endocrinologist, one of the new committee members) did not like your answer on the biochemistry of neurotransmission. They did not flunk you, but they did not pass you either. They said that you would need to schedule a second oral prelim exam sometime within the next few months."

On to the next event.

ON THE ROAD TO DAMASCUS
IN THE NORTH WOODS

*I*tasca, contrary to common lore, is not Ojibway for "place of ravenous insects." As every schoolchild ought to know, it is an acronym derived from Henry Rowe Schoolcraft's designation of the lake bearing its name as the true source (*veritas caput* in Latin) of the Mississippi River, which he discovered after a miserable, meandering, two-month slog through the sloughs and thickets of the various tributaries of the great river in the summer of 1832. (According to apocryphal sources, Schoolcraft swore that his guide, Ozawindib, had taken him past his destination at least five times, like a New York cabby, because he was being paid by the week.)

Kidding aside, the dawn chorus of early summer at Itasca is priceless for the field ornithologist; indeed, one not to be missed by anyone interested in magical natural history experiences. Every morning in the late spring and early summer, beginning in the dark perhaps an hour or so before dawn in the brisk, balsam poplar–scented air, the concert begins—usually opened by the thrushes (robin, Hermit Thrush, Wood Thrush, Veery), then joined by vireos, grosbeaks, tanagers, warblers of twelve species, woodpeckers, chickadees, nuthatches, orioles, flycatchers, cuckoos, wrens, thrashers, sparrows, cardinals, towhees, creepers, buntings, and kinglets (I could hear their high-pitched squeaks back then). This builds to a crescendo of song of exquisite power and beauty a half hour or so after first light, with loons from the lake contributing a haunting continuo, each male of each species proclaiming "I want you; I need you; you can depend on me!" and "I'm the best!" at the top of his lungs (and air sacs).

Lake Itasca, as noted in chapter 5, was home to the field studies campus of the University of Minnesota, where intensive courses were offered during two six-week sessions each summer that focused on field biology (that is, the study of organisms in their natural environment): field botany, field mammalogy, field entomology, and so forth. It is located in the Minnesota portion of the Great North Woods, home of the wendigo: a region of boreal forest and bogs covering vast areas of northern North America and Eurasia. I had attended the first summer session at Itasca in 1973 accompanied by my family. Bon and the kids having departed for western New York, I was on my own this time around, taking a course in field entomology and assisting Dr. Andrew Berger in teaching field ornithology.

Dr. Berger, Andy, was a renowned anatomist and coauthor of the basic ornithological work of the time, *Fundamentals of Ornithology*. He was an excellent classroom instructor as well as a first-rate field man, having discovered the first nests known to science for several Hawaiian honeycreepers. Nevertheless, it had been decades since he had done anything with birds in the North Woods, and he was quite happy to turn the field portion of the class over to me. I, in turn, was delighted to assume this role, having been well prepared by assisting Warner the previous summer. In addition, I had some ideas regarding how to provide students with expanded exposure to birds in the field.

The field course that Warner and I had taught the previous year involved getting the class out from 5:00 a.m. to 7:00 a.m. most mornings during the first couple of weeks to learn the identification of birds by sight and sound. Classroom time was an hour each

Figure 8-1. A Minnesota spruce bog.

day. The rest of the time available to us we used in traveling to various unique habitats in the region to introduce students to their bird communities. With Andy's permission, I added field time spent mist-netting birds, providing students the chance to band them and collect data on subcutaneous fat, molt, sex, and age from a living bird. In addition to the undoubted value to students that the little project could add to the course, I thought that the opportunity to learn something about breeding biology might help me understand other portions of the migrant annual cycle.

The site selected for the mist-netting was La Salle Creek Bog, a two-acre "island" of willow-alder swamp surrounded by forest and lakeshore.

With the assistance of a couple of our most avid course participants, John Barber and Judy Wilson, Mario and I ran twelve mist nets fourteen hours a day for three weeks, accumulating about thirty-five hundred net hours (one net open for one hour) during the last week in June and first two weeks in July.

What we found fascinated me. The following summary is quoted from my book *Bird Migration: A New Understanding*:

> Eight species had breeding territories on the plot, as evidenced by observation of color-banded birds, recapture, and nest discovery. The site was small, a couple of acres, so there appeared to be room for only one or, at most, parts of two territories for each of the species. We confirmed this supposition by observing response of color-marked territorial males using playback of song. Nevertheless, we captured males in our nets far in excess of the number of territories. For instance, we knew from playback that a single Chestnut-sided Warbler pair had a territory on the site, and we knew where the nest for this pair was located. Yet over the course of our netting period during the height of the breeding season, we captured fourteen males. Most of these birds were never seen again after initial capture. The territory owner died on 7 July. At the time his mate was incubating a clutch of three eggs. By 11 July, an unbanded male had taken over defense of the territory (and presumptive co-habitation with the female), continuing there, singing, every day until we left.
>
> There were three noteworthy aspects to these findings. First, the breeding territory for small passerines, which was defined as "exclusive" by most experts, was not. In fact, there were "floater" males [freely] moving through the territory continuously for all eight of the species known

to have territories on the site (many other studies have found this to be true, including one by Darwin). Second, if a territory owner ceased his defense (display and vocalization) for any reason, one of these floater males would take over the territory in short order. Third, female floaters were scarce or non-existent.

The data from this little study provided clear evidence, for me at least, that food was not a critical aspect of the migrant breeding territory, and that females were the arbiters of breeding-period social organization. In fact the contrast for the species that I observed between winter territory (defended against conspecific [members of the same species] intrusion regardless of sex) and breeding territory (basically defended only against males perceived as rivals—at least once mates were incubating) could hardly be more stark.

The attentive reader may be forgiven for wondering at the relevance of this little study to the subject of my thesis research into the nonbreeding biology of songbird migrants. It's a fair question, one that requires some further elucidation of the prevailing theory regarding migrant population dynamics. I have mentioned in earlier chapters that migration behavior was assumed to have evolved over millions of years of gradually increasing pressure to move out of the places and habitats to which they were adapted (their breeding areas) by increasingly harsh winter weather. This movement resulted in nonbreeding-season travel to places where other species of birds already occupied all of the available niches. The visiting migrants could survive in these areas so long as food was superabundant—that is, so long as there was more food available than the resident birds could use. If food was not superabundant, then the migrant was forced to move on in hopes of finding another suitable place with superabundant food. Our first year of work on migrants in Texas and Mexico raised questions concerning the validity of this scenario. We found that individuals of both sexes of many species of migrants were not moving continually throughout the nonbreeding period. Rather, they were staying on the same piece of ground and holding territories for months, apparently from their time of arrival in fall until their departure in spring. In other words, they seemed to have niches that allowed them to compete in these southern environments.

But if these findings are correct, then what is happening on the breeding ground? A corollary of the migrant nonbreeding wanderer hypothesis is that the breeding ground is the only part of the annual cycle where anything important happens. It is the only place

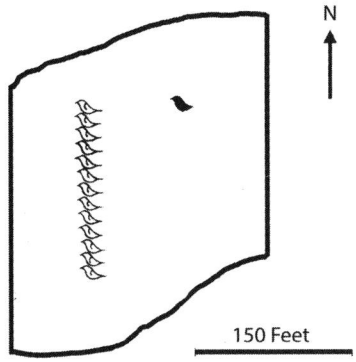

N

Figure 8-2. Male Chestnut-sided Warblers captured over a three-week period on a single territory at La Salle Creek Bog in the summer of 1974. Banded territory owner is represented by a solid silhouette; intruders shown as outlines.

150 Feet

where the migrant has a niche, which is why it can breed there. But clearly, migrants of many species have niches in tropical habitats like rainforest on the wintering ground.

The assumption that migrants have specific niches on the breeding ground is based mainly on the fact that males defend a territory, presumably to secure sufficient food for themselves and their families. But the presence of numerous males passing freely through the territory throughout the breeding period demonstrates that food is not limiting. But if food is not limiting, then there is no need for a niche—the niche being that special ecological space where you (that is, the individual of the species in question) can outcompete members of any other species. Living in a place where you don't have to compete for food is like going to a bird feeder. At the feeder, your niche doesn't matter because there is plenty of food for everybody.

A second issue raised by the Itasca data is the question of the proposed mechanism for migrant survival once they are pushed south each fall by weather. The stated method of survival for these birds once they have left their breeding area is simply to search for food that is not sequestered by birds resident in these southern climes. However, if that were true, one would expect that the winter range of many different migrant species would be the same, or at least quite similar. That this is most certainly not the case is well illustrated by the species we found with breeding territories at La Salle Creek Bog: the Chestnut-sided Warbler and Golden-winged Warbler winter in mixed-species flocks in lowland rainforest in Central America; the White-throated Sparrow in overgrown fields and thickets of the southern United States; the Nashville Warbler in lowland forests of Mexico; and the Veery in the Andean cloud forests of South America.

A third consideration when pondering the role of the breeding range in the migrant life cycle is to recall its relative transience from a geologic perspective. Twenty thousand years ago, Itasca—and everywhere else in northern Minnesota—was buried under several thousand feet of ice and had been for at least thirty thousand years. No habitat suitable for the hundred or so species of migrants currently breeding in the region existed within hundreds of miles. And such glaciations followed by warming had occurred many times previously over the past 2.5 million years. Yet according to calculations made by molecular geneticists, all of the species breeding at La Salle Creek Bog have existed for more than two million years. If this is so, then they all have experienced the complete obliteration of their current breeding range many times in the past.

The Itasca study demonstrated that understanding the nonbreeding portion of the life cycle required an understanding of the breeding portion as well. For this reason, despite the short duration of our work at La Salle Creek Bog, the three-week study there became the first chapter of my thesis.

FALL 1974 IN TEXAS

What's a Girl to Do?

Sex is perhaps the most fraught topic in human discourse. One need only consider the perversion of poor old Saint Paul, where twentieth-century misreadings of his first letter to the Corinthians have converted his advice to his new Greek congregation from a focus on charity to one on romance.

Misunderstandings of relations between the sexes abound, not only in the public in general, but in scientific interpretations regarding those relations in birds as well. A major theme of my recently released volume *Bird Migration* was the realization that male aggression plays no part in interactions between the sexes during the avian breeding season because males lack a penis (except for a few groups, such as waterfowl, as Leda discovered to her dismay) and therefore cannot force sex. So female birds do not have to compete with males during the reproductive period. But what happens when that period is over? This question formed a major part of my research on migrants, and it was during our fieldwork in the fall months of 1974 that we discovered some astounding facts relevant to the issue.

We left Saint Paul early on August 13 and arrived at Welder at about six in the evening on the fifteenth to start the fall field season. By the twentieth, we had nets up again in Hackberry Mott. Although I had no exams for which to prepare, I still found myself strapped for time between trying to run nets and make observations and experiments to test for territoriality among transients. Bonnie helped a lot as always, and I was able to recruit an excellent assistant in Bruce Fall as well. Bruce had a master's from Texas A&M, where he had worked on Nilgai antelope (an introduced African species) on the King Ranch's Norias Division, a vast expanse of oak savanna south of Corpus Christi known as the "wild horse desert." He was doing a contract on plant identification and mapping for Padre Island National Seashore, which brought him out to Welder on occasion to use their excellent herbarium. In addition to his training as a wildlife biologist and plant ecologist, Bruce had a strong interest in and knowledge of birds. In fact, he had published an important paper on new records for Neotropical birds for the Norias Division in the regional journal the *Southwestern Naturalist*. We got to talking in the Welder lab about his work and mine, and he expressed an interest in possibly helping me out, which I was quick to accept. So for a modest stipend and living quarters in the dorm, Bruce came on board in early September.

Bonnie and Bruce ran the nets for the most part while I was doing my little field experiments. I quickly found out that Bruce was not only tireless and dependable but a skeptic to boot—just like all my graduate school buddies up in Minnesota. Skeptics are important in any kind of scientific investigation, but especially so in field biology because so much of the work lacks the clean design of laboratory experiments. When I would come in all enthusiastic about some new observation indicative of transient territoriality, he often would find the weakness in the argument. Such engagement keeps you grounded and on your toes and, as I have since discovered, is good practice for when you get tough questions from the floor during a presentation at a national meeting, where the professional stakes are a good deal higher.

The night of September 13, we got fourteen inches of rain. This time I did not need Dr. Cottam's secretary to tell me that I should get my nets out in a hurry. Even so, nets at the lower (south) end of the mott were already in four feet of water by the time Bruce and I got there just after eight in the morning, with the water rising fast, eventually reaching all but the crowns of the tallest trees.

By the eighteenth, the Aransas was back in its banks and we could start setting nets up again, but the flood had a curious result. It seemed to have scoured the mott of invertebrate life—so much so that migrants recaptured after a day or two almost uniformly showed weight losses rather than gains, as one would normally expect.

Migrants that wintered in Hackberry Mott began arriving again in mid-September, including banded birds from the previous year, Brown Thrashers in particular. These birds often were recaptured in or near the nets in which they had been originally captured the previous winter.

I quickly discovered that wintering-ground arrival is the best time to observe territorial interactions for many species. The only such account that I have ever found in the literature was written by the dean of Neotropical ornithologists, Alexander Skutch. He describes the arrival of Black-and-white Warblers in Costa Rica in October, when there is a good deal of calling, chasing, and fighting for a few days before everyone knows where each individual's territory boundaries are located. This type of interaction was obvious at Welder in several wintering migrant species.

Eric Bolen and his former student, Corpus Christi State professor Brian Chapman, took off on October 14 with Bonnie and me to drive up to the meetings of the American Ornithologists' Union (AOU) in Norman, Oklahoma. This was my first national professional meeting and, as it happened, the most important of my life. My paper, which was scheduled as the last talk on the final day, was to discuss my findings on territoriality in wintering migrants in tropical rainforest.

Bonnie had tried her best to provide me with suitable attire—in this case, a brown leisure suit, the height of fashion in 1974, purchased at a men's store in downtown Sinton. She called this outing "dressing the cat" because she used to go through the same process with an unhappy feline when she was a child and remembered the cat's demeanor as similar to my own. Nowadays, AOS (the organization has been renamed as the American Ornithological Society) presentations are short—ten minutes—with almost no time for questions, and there are several concurrent sessions, ensuring that only those particularly interested in your topic are likely to be there. Back then, we had twenty minutes, and the moderator tried to make sure that at least five minutes were left for questions. Also, there was only one concurrent session rather than the four or five at present, and the paper opposite me was on nest site selection in barn swallows in Mississippi, which meant that I had a chance for a good turnout.

Normally, you would rather have your stuff go out there quietly and unnoticed so that nothing embarrassing happens. But I wanted to get my message to as many of the best ornithologists available to find out what they might have to say. And that's what happened.

When I had finished my presentation, the first question was a tough one from a Minnesota grad student colleague, Joe Wunderle (the botany lab occupant). Joe said that it seemed likely to him that the so-called displays I had mentioned were simply investigative maneuvers common to many species (as Warner had suggested as well). I assured him that the stylized positions taken by territory defenders were, in fact, displays and, in the case of the Wood Thrush, had been described and pictured as such in the famous articles by Bill Dilger on breeding ground territorial behavior of this species.

After the paper and questions were complete, several leaders in the field of migrant ecology came up to question me further at the podium (which could not have happened had I not been the last paper).

All of these illustrious researchers were interested in the evidence of long-term site fidelity and territoriality of migrants in tropical rainforest, especially Dr. Eugene Siller Morton. At that time, Gene was a prominent research scientist at the Smithsonian with several years of experience studying migrant and resident bird behavior in tropical rainforest at the Smithsonian's Tropical Research Institute in Panama. He asked for my address and said that we would be in touch. We were. In fact, this meeting initiated our professional relationship, which has now lasted over forty-five years. (Gene and his wife, Bridget, live in the woods outside Meadeville, Pennsylvania, about a forty-five-minute drive from us here in western New York.)

This opportunity to meet and converse with several of the top investigators in the field of migratory bird biology, however briefly, was important, providing me with the opportunity to have my methods and theories critically reviewed in a public forum and to discover that however renowned, honored, and published a person may be, his thoughts and opinions must meet the test of field study and logic, just like those of anyone else.

Back at Welder, I focused on winter territoriality in Eastern Phoebes, leaving the netting and banding almost entirely to Bonnie and Bruce. Phoebes, which breed across much of the eastern United States, occur only as winter residents at Welder, where they began arriving in mid-October, establishing small territories in woodlands and thorn forests bordering creeks and roads.

I was able to map out territory boundaries using playback for those found along the small tributary running from behind our house a few hundred yards down into Moody Creek. There were seven such territories. Once the boundaries had been mapped, I went out each day and shot the owner. By the time I went out again the following morning, often a new owner had taken over, which was also shot. I continued this regimen for about a month until shortly before our time to leave for Mexico.

The phoebe territories occurred in two different habitat types: mesquite thorn forest and riparian forest borders. The use of these different habitats for territory location by sex for phoebes was not random. I shot fourteen male and four female phoebes holding territories in mesquite and eleven female and nine male birds with territories in riparian forest.

The results of this little experiment were among the most striking of my career. These consisted of four major findings:

1. For migrant phoebes arriving for their five-month stay in South Texas, the possession of a territory is extremely important. Each individual is willing to invest considerable time and effort to obtain and defend that piece of ground.

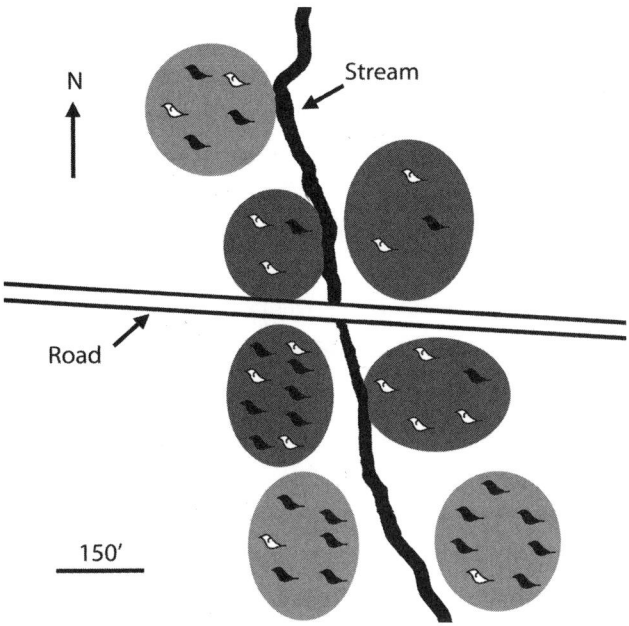

Figure 9-1. Sequence of Eastern Phoebe winter territory holders at Welder in mesquite (*light gray*) and riparian forest (*dark gray*) habitats. Males are black silhouettes; females are white.

2. The sex of the individual does not matter, apparently, in terms of the need for territory. Females must compete with males as well as other females for their own personal, private terrain.

3. In the phoebe and a few other migrants that are territorial in winter, song is used in territorial defense in addition to chip notes. For these species, females sing exactly the same song as males.

4. Although adult male and female phoebes look alike, at least to the human eye, there are small differences in wing length: male wings average 3.30 inches, whereas the female wing averages 3.16 inches. That may not seem like much, but it is evidently enough to make food harvest in riparian forest slightly better for females than males, whereas the reverse is true in thorn forest.

In order to understand these momentous discoveries, one needs to know something about prevailing theories on the reasons for aggression—at least as they stood half a century ago. Back then aggression was considered to be almost entirely a behavior exhibited by males, regardless of the group of organisms considered. In addition, thanks to such prominent theorists as Nobel laureate Konrad Lorenz, the apparent purposes of aggression in humans were often used to explain the behavior in other animals. This perspective persists among many people even today. Jerram Brown, however, made it crystal clear in his seminal paper on the topic that aggression was not solely a male prerogative and that its use in organisms other than people was perfectly rational. In fact, he demonstrated that it was simply a behavioral tool that could be used by any individual of any mobile species to sequester a critical resource for itself. Using a cost-benefit analysis, he showed that aggression, such as that used in the defense of a territory, was based on an assessment of a fine balance between the need for a given resource (such as food or mates) and one's ability to defend it in terms of feasibility and energy required.

A parable may help clarify Brown's meaning. Let us consider the situation of a transient Sanderling named Gretchen. She has been flying south for several hours and on landing is quite hungry. She needs food, which for her consists of tiny invertebrates that live buried in sand washed by the immediate wave line of an ocean beach. If she is alone, no problem; she can simply proceed to eat to her heart's content. However, let us say that shortly after her advent, a second Sanderling arrives whom we will call Jim, who is also hungry and immediately begins to feed on the same small patch of beach as that occupied by Gretchen. Now she has a decision to make. Does she attack Jim in order to defend her piece of beach and its food? Or does she share the piece of beach with Jim? Brown's decision matrix shows the factors that play into her decision. If the invertebrates are widely scattered (low in density), then chasing Jim would be a waste of energy—more costly than the reward. Inversely, if the invertebrates are extremely dense, then chasing Jim would be a waste of energy because time would be better used in harvesting bugs, since there are enough for both of them.

This example illustrates one important factor in Brown's theory of aggression in territory defense—namely, the density of the resource. If the resource is food and it is in low density, then it is not worth defending. Similarly, if the resource is food and it is in high density, then it is not worth defending because there is enough for all. However, if the food is in moderate density—that is, enough for just one individual to harvest in a given time period—then it should be defended.

So density and distribution (in space and time) of the resource are important factors in decisions regarding territorial defense. If the food is important and distributed in a way in which defense against an intruder is feasible, then you should defend it. But the density of potential competitors is important as well. Going back to our example, what if Jim is accompanied by ten additional Sanderlings when he lands on Gretchen's beach? Now what is best? With so many competitors, clearly her energies are best used in harvesting the food as quickly as possible rather than fruitlessly chasing one intruder after another.

The beauty of Brown's theory is that it predicts exactly what has been observed happening in the real world in thousands of species of many different kinds of organisms—birds, mammals, insects, fish, and so on. If food occurs in high density, it is best harvested in groups, whereas if it occurs in moderate density, it may be worthwhile for an individual to fight to sequester if for itself, providing that the number of competitors is small.

If the critical resource is food, then the sex of the individual plays no role in the decision to defend food access, *except* when one's sex affects one's competitive ability, as is apparently the case in phoebes, where slight sexual differences in wing length apparently shift the competitive balance in favor of individuals with longer wings in one habitat type (thorn forest), whereas shorter-winged individuals do better in another (riparian forest).

These findings have great significance in terms of understanding how and where migratory birds spend the nonbreeding period. They also demonstrate the folly of using human behavior to comprehend the meaning of behaviors observed in other animal species.

GAME CHANGE

Death of the Northern Home Theory

No, señor. Lo siento, pero eso no es correcto. Estos no son pájaros de Norte America. Son nuestros pájaros, aunque algunas desaparecen durante la temporada de lluvias.

The quote above comes from a favorite story of Dwain's in which he is out collecting birds in an orange grove in southern Jalisco when he is approached by the landowner, who wants to know what the heck he is up to. So Dwain shows him the Yellow Warbler that he has just shot and explains that he is gathering information on North American birds during their time in Mexico. Whereupon the *campesino* gives the statement shown above, which can be translated as follows: "No, sir. I'm sorry, but that is not right. These are not North American birds. They are our birds, although some disappear during the rainy season."

As for many of Dwain's anecdotes, true or not, this one exactly fits the question at hand. Just how is it that migrants can belong to two different communities thousands of miles apart for large portions of each year? Our first year of work in Texas and Mexico had served mainly to bring the dilemma clearly into focus. Now during the second season at Playa Escondida, I felt it was my job to try to figure this out.

On November 24, Bonnie and the kids and I left Welder, accompanied by Bruce Fall, whom I had talked into going down to Veracruz to help us out during our winter field season in the Tuxtlas. Our vehicle was a small Chevy truck Warner had acquired for us from navy surplus (he had a long-standing relationship with the Office of Naval Research in Minneapolis).

I had built a screened-in, weatherproof structure for the truck bed, which had served as a mobile field lab at Welder and now functioned as a packing crate for all our gear

Figure 10-1. The infamous truck at Playa shown with Virginia Toto, Mario, Jay (in compartment), Bonnie, and Mario's dad.

with a small nest for the kids; Bonnie, Bruce, and I rode up front. Friends with legal expertise who have seen photos of the traveling arrangements provided for the children have commented that such treatment likely is actionable abuse even after the passage of several decades. But I doubt this is true. The kids have shown only passing interest in this possibility in any case.

Arriving at the Brownsville-Matamoros border crossing, we had another adventure. You would think I would have learned my lesson from our escapade at Reynosa, but the hard truth acquired after several decades of fieldwork requiring travel and permits is that dealing with government officials, whether in the United States or anywhere else, often can be problematic, requiring some knowledge that can only be gained through experience. Even then, you have to be flexible and quick on your feet, because each situation has the potential to present unique issues and lengthy delays (or prison). In Mexico, usually an offer of an emolument of some sort will work—or at least it did in the 1970s, '80s, and '90s. I don't know about now. This remuneration is called a *propina* (gratuity) by the recipient and a *mordida* (bite or bribe) by the payer. The knowledge that a simple on-the-spot payment can resolve the situation is enticing but tricky. You have to consider carefully what the situation calls for. How much trouble can the official cause you? How long can they delay you? How much trouble can you cause them? Is bribery part of the culture? If not, as in most of western Europe or the United States (or at least the bribes are well out of reach for normal penny-ante stuff), then you have to try another way (whimper, wheedle, cry, threaten to call their boss or congressperson, etc.). I had an advance tutorial in this subject in November 1989, when Rafa Flores and I drove our research vehicle, a 1979 Chevy Blazer, from Kingsville, Texas, across the entire length of Mexico to our study site in southern Belize. We were stopped ten times on that trip. *Four times* were by customs (Mexico has internal customs stops in addition to those along the border), costing about $1.50 per stop. Each stop goes like this: When the customs (*aduana*) guy approaches your vehicle to inspect, you unobtrusively hand him the equivalent of a buck or two in pesos and say that to save him time, you wanted to let him know that there's nothing to worry about. Then thank him for his help. All four customs stops went as described above—no problem. *Three stops* were by the state police. They are more expensive and less easy to read. I paid the equivalent of $10 at two of the stops, and at the third stop, which included two guys, they wouldn't take a bribe but finally realized that the time it was going to cost them to take us into Veracruz city and wait until Monday (it was Saturday) for government offices to open so that we could purchase a transit license (for taking a vehicle across Mexico from the United States to Belize) was just not worth it for anybody. *One stop* was by a traffic cop in San Fernando, which required open negotiation. He said he was going to arrest us and hold us in jail until Monday, when a magistrate could adjudicate our case. I knew this was bullshit, so I just said, "Fine. Take us in." He said that it could be settled here, but the crime was serious (speeding or something), so maybe $100 and we could be on our way. We said we had no such funds and offered him $10. So it went, back and forth with threats and counterthreats, until we finally settled on $20. The last *two stops* were by the drug arm of the judicial police. These guys were scary. They wore black T-shirts and black jeans and kept Glock pistols jammed in their belts. They looked like normal bandits except for the fancy vehicles. You don't negotiate with them; they stop you, they look in the truck, and if they see something they like, they reach in and take it.

I didn't have all of this knowledge back in the fall of 1974, although I had seen Warner slip the *aduana* officer a couple of bucks when crossing on our spring trip. Anyhow, I tried to play it straight on this crossing. No *mordida*. The result was that the official had us unload every single item for our four-month stay from the back of the truck. It took us three hours to finally pass inspection and move on down the line.

Once we were finally clear of customs, we got another little lesson in foreign relations. In November 1973, Matamoros was a large, sprawling city with few road signs. There was really only one main road south out of town, the Pan-American highway, but getting across the metropolis to that road required a familiarity with the place that I did not have at that time. As we were finishing up loading the truck, a guy who looked like a Mexican American, dressed in an army field jacket, Dallas Cowboys T-shirt, and jeans, came up and asked in unaccented English if we could give him a ride across to the south side of town. I said sure, we'd be happy to. He could show us the way to the Pan-American highway, and we could take him where he wanted to go. When we got to where the road went out of town toward San Fernando, he asked to get out, and we stopped. He then said we owed him $15. Now, I had intended to give him something for his help, but $15 was a ridiculous sum for the service provided in my mind, so I told him no. He said he wouldn't get out until I paid him. I said, "Buckle up, then. I hope you know somebody in Tampico" (four hundred miles south). He then got out in a foul temper saying bad words.

The rest of the ten-day trip south to the Tuxtlas, with a weeklong stop-off in Mexico City to handle social obligations and administrative stuff, was mostly uneventful (couple of blowouts, headlights quit, kids sick), and we arrived at Playa on December 3.

As we began our second field season, I must confess that, for my family, I became more and more like Paul Atreides's sister, Alia, of Dune fame. Only it wasn't Baron Harkonnen who clouded my mind; it was the members of my migratory bird communities in Texas and Mexico. I spent most waking hours (and many sleeping ones) trying to understand what they were up to. As a result, my physical presence often was not indicative of any actual awareness of my surroundings—a disturbing, perhaps even infuriating, situation for those attempting to engage me in normal conversation, or so I've been told.

This second field season was different from the first. Playa was just as spectacularly beautiful, but during the first season, our party constituted the majority of paying customers outside the Christmas holidays. As a result, the patio had been like our personal space, where both the staff and we could relax over refreshments, take in *Bonanza* reruns on the TV (pronounced "bonantha" in the Castilian Spanish dubs), and discuss the day's activities in a convivial atmosphere. Not so during the second season. For whatever reason, Playa seemed to be booked solid most of our time there, so going to the patio was much more like going to a regular restaurant or, worse yet, depending on who the visitors were, to the office. Because the visitors were not solely tourists. They included bird watchers from the United States, France, Germany, and elsewhere, all of whom wanted to talk with us about what they had seen and what they might see; professional overseers of our project from the Instituto de Biología (Gomez-Pompa), Welder (Eric Bolen), and Minnesota (Dwain); friends and family members; and other field biologists (botanists, soil scientists, entomologists, ornithologists). This type of environment placed a pressure on our little community that wasn't there during the first year. Nevertheless, the kids did learn to develop some social skills they might not

Figure 10-2. Jay pointing out a flock of Brown Pelicans coursing over the waves a few yards from shore at the black sand beach.

otherwise have acquired, introducing many of the newcomers to some of the more magnificent vistas and wildlife.

I will mention as a brief aside that I have shown the above photo of my son, who was three years old at the time, in many of my talks and lectures over the years. One such occurred during the summer of 1981, when I spoke at an event at Smith-Wilkes Hall sponsored by the Chautauqua Bird Tree and Garden Club (full disclosure: my mother was the president of that organization at the time). When this photo was thrown up on the giant screen, a plaintive voice could be heard from the audience, keening, "That's not me!" From Jay, of course, who was by then eight years old and not amused at his star turn in front of a hundred or so strangers.

The fieldwork at Playa was different as well. During the first season, I was trying to figure out what questions to ask regarding migrant life in the tropics. The ideas behind the northern home theory as expressed by leading ecologists such as Robert MacArthur and his followers, especially Jim Karr and Ed Willis, were particularly helpful in this regard. They stated that migrants were wanderers away from their breeding grounds, avoiding the use of primary (uncut) habitats such as rainforest, where resident "ecological counterparts" would be present to displace them should they attempt to do so.

Simply setting up mist nets in rainforest to capture and band birds was enough during the first season to establish that the theory had problems. Contrary to its predictions, I had found that a community including several migrant species was present throughout the winter season in the Tuxtla *selva*. However, I had also observed that migrants in disturbed habitats such as pasture and overgrown fields, including several of the species found in rainforest, seemed to behave as wanderers, much as predicted by the northern home theory. I knew that I had to find ways to try to understand the dynamics of this complex system and that mist-netting and banding would not be enough. I would have to spend as much time as possible observing birds in these different environments.

Fortunately, I had major help during this second season with the logistics of capturing, banding, and color-marking of the North American migrant and tropical resident birds. As I have mentioned, Bruce Fall accompanied us down from Welder, and on December 30, he was joined by Chris Barkan (Gene Blacklock's "constant source of irritation" from Goddard College) to help as well. Plus, Bonnie, in addition to watching the kids, doing the laundry, and running her by now famous (to the locals) clinic, continued to serve as data recorder—the key position of responsibility in any field study. When the guys came back from running the mist nets, it was Bonnie who oversaw the banding, color-marking, and data collection (date, time, species, band number, color-band sequence, fat, molt, weight, sex, age, net number) and entered all of the information into the catalogs (a main catalog and a species catalog so that an appropriate color-band sequence could quickly be determined). I shared mist-netting duties, of course, but spent as much time as possible, usually five or six hours each day, in a blind watching my birds.

The first field season involved netting only in *selva* on Raul's property near the *cabañas*. For the second season, I decided that I needed to net in a second *selva* site as well as sites in one- and five-year-old second growth. The addition of these was physically challenging, especially the second *selva* site, which required a twenty-minute climb up a shoulder of Cerro Balzapote and was a nightmare of massive trees, vines, tangles, and arroyos on a sharp incline.

It was while setting this site up that I suffered a serious turn. Snakebite and murder (parochial matters) were the two chief causes of death for adult male peasants in the area, and it was while clearing a net lane for this new forest site that I came close to cause #1. The spot chosen for net 5 was steep. I walked down it, flailing at the main intervening branches with my machete as I went, intending to clear the undergrowth, which was

Figure 10-3. Chris Barkan threatening Bonnie with a bug. She was unimpressed.

Figure 10-4. Bon and Virginia Toto admiring a fer-de-lance, former resident of net lane 5 at my new *selva* site, February 1975.

quite dense. It was during my return when I suddenly realized that due to the steepness of the site, I was face-to-face with a coiled, six-foot fer-de-lance. Somehow I had stepped over it, rather than on it, on my way down. All's well that ends well, and it is now safely ensconced in the Minnesota herp collection. Mario, and probably all of the rest of our field-workers whether they knew it or not, including Bonnie (in Hackberry Mott), had had similar experiences—fortunately, without dire consequences.

Once the sites had been set up with all of the mist nets placed in a grid (one net every one hundred feet), I turned most of the net-running over to Bruce and Chris. A normal day would begin at dawn (about six in the morning), when I would go out and set the nets on the site or sites that I wanted them to run. After breakfast, I would usually help with the first net check. Then I would head out to do some sort of stakeout experiment on the territory of one or more of my color-banded migrants. These stakeouts assumed several different forms, depending on the species being tested and the birds that had been captured during the first net run. Usually, I would take a bird from that run—say, a Hooded Warbler male—and place it in a cage built of mist-netting somewhere on the territory of one of my color-banded hoodeds. Sometimes I would accompany the stakeout with playback of a recorded vocalization, usually a chip note but sometimes song if the species was known to use song on winter territory. Then I would get into my blind, a burlap structure with an opening for me to see through and take pictures from. I didn't always use the cage. Sometimes the bird was tied to a board; sometimes a stuffed bird (mount) was used; sometimes it was accompanied by vocalization, sometimes not; and on a few occasions, I used a mirror to confront the territory owner with itself.

It is somewhat difficult to explain, but the bird communities on my study sites became my world. The more time I spent with them, the more I came to know how they spent their days because while I was busy conducting my experiments, the normal life of the forest was going on around me. The focus of each experiment was always a color-banded bird well known to me. I knew these birds' history well—date first captured, territory boundaries, sex, age, and weight—and I thought about what they were up to constantly, even when I wasn't in the blind. I'm afraid my marriage suffered as a result.

The most striking discovery of the second season was the rate of return of migrants banded the previous year. Other researchers had reported returns of North American migrants to tropical sites where they had wintered previously, but none had mapped the territory boundaries for each individual prior to departure in order to test the precision and rate of return for the next winter season. As described above, we had done this for the Hooded Warbler, mapping ten rainforest territories during the 1973–74 season. When we arrived in December to commence the 1974–75 season, one of our first activities was to capture (or recapture) all of the Hooded Warblers on the site and map the boundaries of their territories. This process took us about a month of intensive mist-netting, playback of territorial chips, and "pushing" using the flush technique as earlier described in order to get all of the resident territory owners color-banded and find out exactly where the boundaries of their territories were located. The results were astounding. Six out of the ten birds that had held territories on the site in 1973–74 had *returned* to reoccupy those same territories in 1974–75, and the boundaries of those territories were roughly the same as they had been the previous year.

This finding was certainly the most significant of my entire doctoral study—and perhaps of my career. Not only does it demonstrate conclusively that these migrant Hooded Warblers were not wandering during the wintering period; it implies two additional factors of extreme importance. First, for 60 percent of Hooded Warblers wintering on our study site, which is likely a number equal to those surviving the round trip, a winter territory is critical to their survival. Second, these migrants hold niches in their tropical wintering environments, which is completely contrary to the northern home theory of migrants as interlopers in the bird communities in which they spend the winter months. Furthermore, the phenomenon of returning to rainforest winter quarters in subsequent years was not restricted to Hooded Warblers. We had returns for all of the common migrant species of our rainforest community. It is true that the return rates for these other species were far lower than for Hooded Warblers, but we had made no special effort to capture and band every territorial individual or map their territory boundaries during the 1973–74 season for any species other than hoodeds. In addition, we had not tried to find all birds banded during the previous year for any migrant other than hoodeds during the 1974–75 season. Had we made comparable efforts for the other territorial migrants wintering on our *selva* study plot, I have no doubt that we would have found similar rates of return.

As a result, we were able to confirm that there was a community of North American migrant species occupying and showing both short-term (within a given winter season) and long-term (from one year to the next) site fidelity in the iconic primary habitat of the tropics: rainforest.

Individuals of nearly all species of migrants that wintered commonly in rainforest habitat showed some evidence of territorial defense, including regular (that is, hourly) use of vocalization (song or call) as advertisement, response (approach) to playback of vocalizations, and use of stylized display, attack, and chase of "intruders" (free flying, caged, tethered, or mounts) of the same species. However, the strength of the response to intruders varied markedly among species, and even from individual to individual of the same species. Hooded Warblers, for instance, could be depended on to approach in reaction to initial playback or first-time introduction of a caged bird or mount about half of the time, and once "sensitized" (that is, after finding an "intruder"), they almost always approached, displayed, and attacked on subsequent exposure. Ovenbirds and Kentucky Warblers were far less aggressive to artificial intruders (caged birds or mounts), although they responded immediately to free-flying intruders.

The long hours spent in the blind revealed some notable insights into the curious functioning of the system, mostly from work with Hooded Warbler territory holders, but also from observations of Wood Thrush, Ovenbird, and Magnolia and Kentucky Warblers. As we had found for territories on the breeding ground, wintering-ground territories were not exclusive. Floaters, evidently nonterritorial individuals, would be seen foraging occasionally right through the space defended by a known (color-marked) territory holder. The reaction by the territory holder was fascinating. So long as the intruder made no vocalization and stayed only for a short period, it was ignored, but a chip note or some change in posture by the invader would bring on an immediate attack and chase response. I had noticed a similar response to artificially introduced (caged or tethered) intruders as well. So long as they were quiet and still, they were generally ignored by the owner. Only when they chipped or when I used the playback of a chip to accompany their "intrusion" did the owner pay any attention. The origin of free-flying intruders was unknown for the most part. However, in two cases they were known birds that had been captured (and color-marked) on neighboring second-growth sites.

I did more work with the Hooded Warbler than with any other migrant found wintering in *selva* because of its abundance, visibility, and ease of capture. Visual display patterns used in the defense of territory are probably no more varied in this species than in any of the others known to defend winter territories in the Tuxtlas, but hoodeds responded more readily to artificially introduced intruders so that more complete descriptions and films of displays could be obtained.

Direct attacks and bill snapping are also included in the repertoire. Displays seem to exploit the gold cheek patch framed by the black hood and bib, yet fewer than 5 percent of females possess this pattern and are perfectly capable of defending territory against male intruders.

None of the territorial birds died of natural causes during either the 1973–74 or the 1974–75 field seasons. In February 1975, Raul had all of the forest south of the road (about two acres) cleared as pasture, which served as a natural experiment for us. Many people seem to think that territory defense in winter is like a little game played by birds to pass the time. Raul's destruction of the habitat on the territories of these four birds served as a sort of ad hoc experiment. If it was just a pastime, then one might expect the birds to simply move into the forest in the immediate proximity, forcing the hoodeds already there to just shove over a little. However, that is not what happened. Of the four birds with territories in that area, two were later recaptured as floaters in the forested area north of the road; the others simply disappeared. The two floaters ultimately disappeared as well after having been caught on the territories of a couple of neighbors over a few days.

For me, this finding further illustrates how important the territory is to its holder. If the habitat on that territory is destroyed, the bird is no longer sedentary. Now it is a wanderer, and it has to move to find the food necessary to survive the winter, which had been provided by the now destroyed territory.

Once territory was established upon arrival in the fall (or upon the replacement of a removed territory holder), the owner showed a high degree of sedentariness, seldom moving beyond its boundaries throughout the winter season. In over seven thousand net hours on our *selva* site, only twice were territorial birds captured more than 150 feet from their original point of capture, with the exception of the two previously mentioned birds whose territories were destroyed. Tests with caged intruders and mounts showed that winter territories likely were defended continuously from the time of establishment in the fall until the time of departure in the spring.

To test the relative importance of vocalization, color patterns, and movement in stimulating territory owner response to intruders, forty-five separate experiments were performed in which different combinations of live birds, played-back vocalizations, and stuffed birds were placed in the territory of a marked bird. The results present a fairly clear pattern of the aggressive territorial response to intruders. An intruder sitting still is unlikely to be seen, let alone attacked. If it is seen, it will be approached and investigated, possibly displayed to and possibly attacked. A vocalization by an intruder is usually investigated, and if a bird of the "right" color pattern is found, it is usually displayed to and attacked, often repeatedly. On a few occasions, sessions went more than three hours with a territorial bird attacking a caged bird or stuffed mount in approximately one-minute intervals throughout that time. Once a territory owner had seen a mount or caged bird and heard an accompanying vocalization, it often would continue attacking and displaying whether the intruder stopped vocalizing or not.

In several cases where known territorial Hooded Warblers were removed from their territories, they were replaced by birds that had not previously been captured. Once established on their new territory, they became sedentary and aggressive toward conspecific intruders, whether artificially introduced or free flying.

"Caged intruder" and stakeout experiments, like those described above for the Hooded Warbler, were performed on nine other migrant species. For all species studied in this way, color patterns, sound, and behavior were found to be the important factors in triggering a stylized, aggressive response, although the relative importance of these factors varied considerably from species to species.

Work in the overgrown pasture and second-growth habitats yielded another interesting finding. Individuals of several of the same migrant species found in forest were also found in these habitats where their behavior often was quite different from their *selva*-inhabiting conspecifics. Banding data revealed few recaptures, and there was little evidence of territorial defense. In other words, many of these birds seemed to be wanderers, as predicted by the northern home theory.

The sex ratio in the various habitats in which migrants were found wintering varied markedly. For instance, in the Hooded Warbler, of twelve birds known to hold territory in *selva* in 1975, ten were males. Of two birds holding territories in older second growth, both were females. In general, males captured in rainforest outnumbered females eight to one, while females slightly outnumbered males in second growth. These ratios are based on specimen data as well as observations so that the skew is not caused by females with male-like plumage, which, in any case, composes a small percentage of the female population of Hooded Warblers according to specimens in the Minnesota collection. During migration, both sexes were captured in approximately equal numbers.

The Hooded Warbler was not the only species in which the sex ratio was found to be skewed. Males outnumbered females captured in rainforest, whereas females outnumbered males in second growth for Wood Thrush, Wilson's Warbler, and Kentucky Warbler, while females outnumbered males in the Black-and-white Warbler.

I have described documenting by banding and the observation of color-banded birds the long-term (seasonal) presence of twenty species of migratory birds on my *selva* study sites; another twenty species or so of migrants were present in neighboring sections of other major habitats, such as wetlands, pasture, and both younger and older second growth. I watched carefully for evidence (similar foraging behavior or active display and attack) of tropical resident species that might qualify as ecological counterparts for the migrants wintering in these environments and found none. Of course, the acid test for the presence of ecological counterparts for migrants would be migrant absence as long-term residents (weeks or months) from tropical habitats, which the netting, banding, and observations over two field seasons demonstrated was not the case. The import of all of these observations and findings was quite clear, even after just a few months—namely, that many members of the migratory bird species wintering in the forests and fields around Playa and the UNAM biology station were not wandering interlopers. For the period of their stay in the Tuxtlas, they were card-carrying members of the avian community.

Eric Bolen brought me my mail from Welder during his January visit to Playa, which included, among other things, a letter from Dr. Fred Gehlbach, esteemed professor of zoology at Baylor University. I have mentioned that the opportunity of presenting findings before peers at a national meeting is an important testing ground for ideas; another is to publish them. The publishing process allows you to place your work before selected experts (the editor and his or her chosen reviewers) and then, if accepted by the journal for publication, before the academic community at large. Eric was the first to provide me with this opportunity. After visiting our Mexican study site in early 1974 with Dr. Cottam, he became excited about our work and suggested that I submit a paper to the *Bulletin* of the Texas Ornithological Society, which was edited by a good friend of his, Dr. Kurt Rylander. The bulletin was a regional journal with a small readership of mostly nonprofessional bird watchers. Nonetheless, like my paper at the American Ornithologists' Union (AOU) meetings, it had the potential to give me a chance to put my ideas out there, and I jumped at it. Dr. Rylander eventually accepted a revised version of my short article, and the paper came out in the December 1974 issue. This paper provided a brief summary of my first year of findings for Texas and Mexico regarding migrant behavior in their tropical and subtropical wintering homes and proposed that if such habitats continued to be destroyed at rates reported by environmental agencies, populations of migrants dependent upon them would suffer.

There were a couple of positive results from this effort. First, it gave me my first publication. Quantity and quality of publications are the basic measures of accomplishment for any research scientist. Second, it stimulated a long appraisal from Dr. Gehlbach. His letter was highly critical of my Mexican findings, citing in particular the work of Elliot Tramer, University of Toledo biology professor (and MacArthur disciple). Dr. Tramer's paper addressing the same topic of migrant wintering ecology as my own, and also based on work in Mexico (Yucatán), reached conclusions quite different from mine. He found that migrants were among the most common inhabitants of gardens and residential areas of Merida but were also common in the native dry forest that covers much of the northern part of the peninsula. He noted that migrants were seldom or never found in stable tropical habitats like rainforest. Dr. Gehlbach averred that certainly, I must be mistaken in my findings, which were obviously contrary to Dr. Tramer's as well as most previously published information on wintering migrants in the tropics.

Initially dismayed, I eventually realized that if my findings weren't important, a person as prominent in his field as Dr. Gehlbach would not have bothered to critique them. Also, I found that I was able to rebut his criticism using data from our Mexico work. But perhaps most importantly, Gehlbach's letter underscored a fundamental conundrum of wintering migrant biology—namely, how could both Tramer and I be right? How could wintering migrants of the same species behave in one way in residential areas, pastures, and second growth (small, loose flocks of wanderers) and another way in primary forest (solitary territorial individuals showing long-term site fidelity)? I had an inkling back in early 1975, but it took me many years of field study, collaboration, and discussion with experts on migration from other parts of the world before I was able to develop a hypothesis that could reconcile these seemingly inconsistent sets of observations.

THE ROSETTA STONE

Spring 1975

Then was I like some watcher of the skies
when some new planet swims within his ken.
—JOHN KEATS, *ON LOOKING INTO CHAPMAN'S HOMER*

We left Playa Escondida on March 17 and arrived at Welder on the eighteenth. I got the net lanes cleared and nets back up by the end of the month with help from Chris Barkan, who had volunteered to assist me again through the spring season. Reconnaissance revealed a notable change. Although the mott was mainly riparian forest, a swale now extended down the middle of it, perhaps two hundred feet west of the Aransas River banks. This area had been mostly dry when we began work in the fall of 1973, but after the deluges that occurred in October 1973 and September 1974, it formed a large temporary pond bordered by extensive muddy banks, twenty to thirty feet in width. These mudflats had not existed during previous field seasons, but they provided not only a perfect foraging habitat for transient Northern Waterthrushes but a phenomenal opportunity to observe, capture, and take data (daily fat content and weight) on several of them during their stopover at the mott.

When I realized this, I decided to turn over all of the netting duties to Bonnie, Chris, and Bonnie's friend Karilyn Mock while I would focus on the fabulous opportunity presented by the mott's temporary pond. From April 10 until May 16, I had ten nets, with one net every fifty feet or so around the periphery of the pond, each extending from the water's edge across the mud toward the forest. These nets were run only to catch newly arrived waterthrushes or recapture specific waterthrushes at specific times. A total of 122 waterthrushes were captured, weighed, and color-banded during this period. As closely as possible to twenty-four hours after initial capture, I recaptured waterthrushes frequenting the pond edge, recorded weight and fat content, and released them, repeating the process every day until the bird left the area.

Figure 11-1. A portion of the temporary pond in the middle of Hackberry Mott showing mist nets in place to capture Northern Waterthrushes on their mudflat territories, April 1975.

A deer blind atop a fifteen-foot stand was put up on the east shore, from which it was possible to survey the entire pond and its muddy banks. I spent a total of sixty-two hours in that blind over the five-week period, recording interactions between individual waterthrushes—including displays, vocalizations, and chases—and constructing daily territory maps for each individual. This combination of data collection circumstances made the temporary pond a perfect Rosetta Stone, enabling me to relate the behaviors observed in the birds to their probable physiological state.

Jerram Brown defines *territory* as "a fixed area from which intruders are excluded by some combination of advertisement (such as scent or song), threat, and attack." The existence of a fixed area actively defended and advertised is strong evidence that the social system is based on territoriality. Transient Northern Waterthrushes, stopping to rebuild fat reserves at Welder, exhibited this type of system. Males and females held separate territories, which each individual defended against intrusion by members of either sex. The primary means of defense was advertisement, for which the "chink" call note was used by territory holders an average of eighty-six seconds per hour. Territorial neighbors did not normally cross each other's boundaries, so most hostile activity was directed toward nonterritorial intruders. Attacks on these interlopers were frequent and usually turned into chases when the intruder fled. If an intruder did not leave when it was attacked, a fight ensued that involved physical contact. When two birds were foraging on the ground and happened to confront each other, a visual display was often given by one or both birds. Territories were held an average of 3.2 days. Waterthrushes captured anywhere in the mott were released at the pond. Twenty of the 122 captured were recaptured at the pond 1 day or more after release. The average weight change for these birds was +0.4 grams. Six birds were recaptured away from the pond area, for which the average weight change was +0.3 grams.

As these experiments show, individual transient Northern Waterthrushes used a territorial system to guarantee critical resources for fattening. The average length of stay for territorial birds was 5.6 days, but the average time to hold a territory was only 3.2 days. I interpreted this to mean that in most cases, at least 2 days passed before a bird could gain a territory. Whether or not a bird was in possession of a territory had profound effects on its weight: in the days prior to obtaining a territory, birds showed weight losses, while they showed significant daily weight gains after obtaining a territory.

This work at Hackberry Mott with waterthrushes was, as I have stated, the Rosetta Stone of transient biology. The chance to observe and measure the relationship among behavior, physiology, and environment is vanishingly rare in field biology, but there it was. A unique set of circumstances combined to allow me the opportunity to do this work in the spring of 1975. These experiments—which, to my knowledge, have never been duplicated or even approached—provide a striking insight into the behavior and physiology of free-flying transient birds and set the stage for what should have been follow-up experiments in the lab, where details of hormonal and other physiological changes and their associated behaviors could be assessed, manipulated, and tested under controlled conditions.

I tried for forty years to find the funds to do this work (seven proposals to the Smithsonian Scholarly Studies program, three to the NSF) but was unable to do so, despite having the ideal setup at my lab at the Smithsonian's Conservation and Research Center—with cages and trained animal keepers available on-site and a willing collaborator (Steve Monfort) with world-class credentials in the measurement of hormones in feces and urine, which is critically important because no handling of the bird is necessary (manipulating a bird to take blood for hormonal measurements causes stress, which changes hormonal profiles).

The reader may wonder why I failed, to which I can only answer that my proposals could not survive the review process. No sense whining about it; it just didn't happen. In any event, as a result, we still have a poor understanding of these linkages. Science, however, is patient. Eventually, someone will follow up, and the critical ties among behavior, physiology, and environment for transient birds will be clearly understood.

I have mentioned that I was not good company in Mexico during the weeks when I was trying my best to think like a Hooded Warbler. I was worse in Texas when trying to think like a Northern Waterthrush. Finally, by May 20, Bonnie had had enough. Bird brains may be interesting, but they make for poor company in the long run. She took the kids and left for her mother's in Jamestown (New York), while I took a few more days to finish up at Welder before heading off for Minnesota.

A WORKING HYPOTHESIS

Bonnie and I were able to patch things up once I had emerged from my virtual immersion in the world of migrant behavioral ecology, and we settled back into our third-floor apartment at 672 Summit Avenue for a year that was to be spent mostly by finishing up my coursework, taking my prelims (again!), writing my thesis and getting it accepted by my graduate committee, and looking for jobs.

My PhD committee had loved my written prelims. However, my reconstituted committee had not liked my oral prelims as I described in chapter 7, forcing me into a highly unusual do-over after I had completed my thesis research in the summer of 1975. This second exam was once again in the windowless classroom off from the museum's Touch and See space, and it went worse than my previous orals. As before, the committee was new, three of the five members having been replaced. Frank McKinney was one of these replacements.

McKinney was probably the most famous member of the Bell Museum faculty (in the field biology world). He had earned his PhD at Oxford under the tutelage of Nobel laureate Nikolaas Tinbergen. Prior to the exam, I had talked with Frank, from whom I had taken a course in animal behavior (receiving an A), about what he thought I might read to prepare, as I did for all of my committee members. He said that I should be sure to read E. O. Wilson's recently published book *Sociobiology*. Well, I knew that. The entire world was talking about that volume at the time because of its implications regarding human behavior.

During the exam, however, McKinney asked only one question: "What is a 'leading line'?" Now, anyone who has studied migration from a duck perspective should know the answer to that. McKinney's fame was based on his PhD thesis, which employed the behavioral characteristics of pond ducks, like the Mallard, in their taxonomic classification. In his view, any good student of the phenomenon should know this characteristic behavior of migrating waterfowl, which tend to form a long line of birds during migratory flight along any impediment or barrier, like the shore of an ocean or large lake. I, however, studying the ecology of migrating songbirds, had never heard of the term, and said so. (As a note of interest, the term *leading line* occurs nowhere in Wilson's seven-hundred-page classic.)

The other question that I completely flubbed during this second go-round was from Bill Schmidt, the animal physiologist on the vertebrate zoology faculty (and from whom I had also received an A in his course). He asked me to explain the binomial distribution. Now there really is no excuse for a PhD candidate not to know the answer to this question. The binomial distribution is the fundamental theory underlying parametric statistics. That I knew, but I did a poor job explaining the theorem.

In my memory, those were my only two problems in the exam, but the result was similar to my first go-round. The committee did not verify that I had passed my oral prelim. Instead, they said that my final oral would serve as my oral prelim.

To say that this is unusual is a gross understatement. *Unprecedented* would be more appropriate. The "final oral" is generally a victory lap. It is given after the thesis has been accepted by the committee. At Minnesota, the procedure was for the candidate to give a public seminar on his thesis, open to anyone in the university community—or, indeed, the community at large—to attend. After the seminar, the candidate and committee would meet for the final oral, which normally might involve a few softball questions about the candidate's thesis, with a final singing-off on approval of granting of the PhD for the graduate school.

Not so fast, my friend. My final oral was supposedly to be like an oral prelim. In other words, the committee could still decide not to grant me my degree. As a disinterested observer, you, the reader, may feel that the failure to award the degree at this stage would be a highly unlikely prospect. Maybe so, but I can tell you that that was not how it felt for me.

Once again, my committee had been rearranged. I had two new members, one of which was my old nemesis, Bud Tordoff, who had replaced McKinney, and after completing my seminar to the faculty and graduate students of the Bell Museum, it was Tordoff who led off the questioning for my in camera session with my committee. His question was "The feather shafts of individual birds occasionally are found to have holes located at the same precise distance from the bird's body. How can this be?"

The answer, of course, is obvious—if your brain is not completely frozen in terror at the situation. The bird was infected with feather mites, which chew into the shaft in the nestling stage, at the age of four or five days, when all body feather shafts were half grown. Pretty simple, which was the answer I gave.

Tordoff smirked and said, "I guess this guy is ready to get his doctorate."

This question infuriated me. Here I was teetering on the precipice—my career in the balance—and I am given the field biology equivalent of an idiot test, like "Where was Moses when the lights went out?" or "Who is buried in Grant's tomb?" Oh well. I did get the degree in June 1976.

Why did I have so much trouble? Now that I have had some years in the business, I have some theories. The first has to do with my relatively weak academic background. Without Warner's sponsorship, it is unlikely that I would have been admitted to the Bell Museum program, even before Tordoff's arrival. The second had to do with my advisor. When Warner took me on, he was arguably the most powerful professor in the department. However, after my postwar entrance, that was no longer the case. In fact, he was on the verge of dismissal, a rare situation for a tenured senior professor. Based on my experience on other committees, the students of powerful senior professors do not have trouble with their committees. Any committee member junior to the advisor would know full well that if he or she caused any problems for the senior member's student, they could expect significant reciprocity. No way would they court such a possibility. Warner, however, had lost all seniority and was fair game.

These observations help explain why my committee might have failed me. However, they do not clear up why they did not fail me. This circumstance, I think, requires some serious speculation, helped perhaps by a conversation I had some years later with one of my committee members who had become a personal friend. He said that the committee, at least some members, felt that I was arrogant.

Now this seemed absurd to me. How could they consider a scrawny, diminutive, ignorant, trembling novice to be arrogant? But I have had a few decades since to think about it, and I think I know what he meant. The first problem for the committee, regardless of their opinion of me or my advisor, was that by the time I took the oral prelim, I had full funding for my thesis work—not only for me but for two additional students in the department. Furthermore, that funding had been granted to me personally, not my advisor. What would happen to the grant if they kicked me out? The second problem was that I had already taken the written prelim and done well, based on the judgment of my committee, some of whom were no longer members. The third problem was that the oral examination was in no way preliminary. I already had taken several courses in the department, achieving high marks in all of them, and had a year of fieldwork under my belt. Therefore, my circumstances as I took my first oral prelim in June 1974 may have given members a sense that I might feel that I did not need their approval (which was, of course, absurd). This situation was exacerbated by the time I took my second oral prelim in June 1975 by the fact that two of the three chapters of my thesis had already been accepted for publication, a factor that likely upset my committee a lot.

I had not even thought about pissing off my committee when considering the pre-publication of stuff that was to appear in my thesis. My concern was to get as much of my work out there as quickly as possible because I knew that publications were how you got jobs. A PhD is only a credential. Without publications, it means little, at least at that time in my field. Besides, the whole idea of trying to write up chapters as publications was, once again, serendipity, as I explain below.

It was while I was working at Welder in the spring of 1975 that Fred and Fran Hamerstrom stopped by. They were friends of Warner's and had spent a week or so at Playa Escondida a month previously, trapping and banding raptors.

In addition to spending time together on the patio at Playa, Bonnie and I had met them during earlier visits by them to Welder. We considered them personal friends, and

Figure 12-1. Bonnie, Jay, Brigetta, Mario, Fred Hamerstrom, and Dwain admiring Fran Hamer-strom with her White Hawk capture at Playa. The famously pungent (roadkill on the roof) Hamerstrom cadaver van is to Fran's left.

as former students of legendary wildlife biologist Aldo Leopold and famous research scientists in their own right, mostly based on prairie chicken work at Wisconsin, I thought that their presence for a week at Welder presented a golden opportunity. I felt that what Mario and I had found during our work on migrants during the breeding season at Itasca raised some interesting questions, perhaps worth publishing, and that the Hamerstroms could provide valuable advice and insight into how to present the work in the most publishable way. Accordingly, I quickly wrote up the paper and presented it to them for review. Unfortunately, they didn't like it and said it was unpublishable for a number of reasons. Fran, in particular, sat me down to break the bad news in the Welder coffee room and explain that the work really had no scientific value. Fortunately, I did not believe her. I took the specific comments to heart, rewrote the paper, and sent if off to *American Midland Naturalist*, a regional journal of excellent reputation published by Notre Dame, where it was quickly accepted after review. Originally, I had not intended that the breeding-ground work would make up a part of the thesis, but once it had been accepted for publication, it seemed to me that it could fit perfectly as part of an examination of the entire migrant life cycle.

The publication of the second part was equally serendipitous. It was in June 1975, after our return to Minnesota from our Texas and Mexico field seasons, when the German editor of *Oecologia* stopped at the university on a lecture tour. At the usual post-lecture conclave, he and Warner got to talking about our migrant work, and he invited us to submit a paper on it to his journal. *Oecologia* was, and is, one of the more prestigious international journals in the field of ecology, so we readily accepted his invitation. I wrote up the paper summarizing the Northern Waterthrush findings from our spring 1975 efforts, and it was accepted for publication by December or so of 1975. So you see, I had no intention of dissing my committee by getting chapters of my thesis published before they had approved them. It just worked out that way.

Now, you might think that such a situation would make the committee happy. After all, a student of theirs was already doing well in the larger academic environment beyond the university. Think again. The outside submission of thesis material before the committee has had a chance to rule on their academic value is an abrogation of their prerogative. This concept is obvious to me in retrospect. In fact, many institutions prohibit the publication of thesis chapters prior to committee approval for this exact reason. I personally had no notion of this problem, and Warner certainly did not advise me of its existence. All I knew was that it was a difficult work environment out there, and my degree only gave me standing for application for jobs. Publications were what made you competitive. So I did my best to make sure that I had some when it came time to apply for positions upon graduation. I did not understand that I was tying my committee's hands regarding the judgment of the value of my thesis.

These factors, I think, were the source of the "arrogance" perception. How could they fail me after I had already been accepted by their colleagues and the granting and publication communities? They couldn't. But they could make me suffer. And they did.

Despite marital and committee problems, I found the time spent writing up my thesis to be exhilarating. The more background information I read, the more excited I became. The breeding and transient work opened vast vistas of new understandings of migrant life cycle ecology for me that seemed to offer a lifetime of research opportunities. But the key part was the wintering-ground findings. These, I felt, would change how the scientific world thought about migrants. Because I knew I had the data for a critical evaluation of the northern home paradigm.

According to the northern home theory for the evolution of migration, migratory birds that breed in temperate and boreal portions of the Northern Hemisphere originated as residents of those regions who evolved migration to south-temperate and tropical areas over a period of millennia in response to increasingly longer periods of cold weather during winter months. If this theory is correct, then the following must also be true:

1. **A migrant's niche—that is, the ecological space including food and habitat unique for its species—exists only on its breeding grounds.** The fact that we found twenty species of North American migrants, one-third of the total of common passerine species in the *selva* community, to be resident on small plots of ground throughout the winter season indicated that the niche for at least some species of migrants was present in tropical rainforest.

2. **Once the migrant leaves its breeding grounds, it is outside its niche, which means that it must survive as a wanderer (fugitive species).** Members of the twenty species of the Tuxtla *selva* community demonstrated both within-season and between-season (from one winter season to the next) site fidelity on our rainforest study plots. They were not wanderers. This phenomenon was most clearly demonstrated by the Hooded Warbler, for which we were able to map the *selva* winter territories for ten marked birds. All ten birds remained on those territories throughout the 1973–74 winter season, and six of the ten returned to those same territories in fall 1974, where they

remained throughout the 1974–75 season, or at least most of them did. Raul cut the forest on which four of the territories were located in February 1975, after which two of the birds disappeared immediately and two others moved over into neighboring uncut forest already occupied by territorial conspecifics, where they were captured and/or seen for a few days postcutting until they disappeared as well.

3. **Stable environments in areas outside the breeding grounds will be occupied by indigenous species, and in these environments, species similar to the migrants will exist as "ecological counterparts" and fill niches comparable to those filled by the migrants on their breeding grounds.** The leading theorist on migration predicted that indigenous resident bird species in the tropics would fill the niches occupied by migrants on their temperate breeding grounds, preventing migrants from long-term residency in most major tropical habitats. Therefore, an important part of my fieldwork in both Texas and Mexico was to spend many hours each day observing individuals of the various species of migrants that we found living in the principal habitats of our study areas to see if there was evidence of members of resident indigenous species similar in foraging habits to migrants and/or aggressive toward them. As luck would have it, the textbook example of the ecological counterpart involved the Hooded Warbler, the star performer on my Playa study sites.

MacArthur acolyte E. O. Willis had published an important paper relevant to the "ecological counterpart" issue in 1966 in the journal *The Living Bird*, in which he described interactions between resident Spotted Antbirds and wintering migrant Hooded Warblers in Belize rainforest. The antbirds, as indicated by their name, appear dependent on colonies of army ants for their food. Antbird pairs, male and female, defend a space around the ant colony as it moves through the forest, snatching up invertebrates as they attempt to escape the swarm. Willis observed that Hooded Warblers occasionally attended such swarms as well, from which they were often driven by antbirds. Based on this observation, he concluded that the Spotted Antbird likely was an ecological counterpart of the Hooded Warbler. The case collapses in the Tuxtlas despite the presence of the same army ant species as that found in Belize (*Eciton burchellii*) and an antbird, the Black-faced Ant Thrush (but no Spotted Antbirds). As discussed above, I found that Hooded Warblers resided on small individual territories in forest throughout the winter period. Certainly, they took advantage of the prey dislodged by the ants when swarms crossed into their territories, but I saw no interactions between hoodeds and any tropical resident species during such times, let alone the ant thrush, which was seldom seen. The closest thing to an indigenous tropical forest counterpart for the migrant hooded on my study sites was the Sulphur-rumped Flycatcher, a bird comparable in size and foraging behavior but different in microhabitat use (thickets versus open understory).

Similar kinds of analyses were performed for all other migrant members of the Tuxtla *selva* community, and they too were found to show short- (seasonal) and long-term (annual) site fidelity in tropical rainforest. No evidence for the presence of indigenous "ecological counterparts" for migrants was found. In fact, detailed notes on foraging behavior revealed that the species most comparable were other migrants, such as the Wilson's Warbler and Magnolia Warbler, which forage at roughly the same height on average in the forest canopy (about sixty feet), and the Ovenbird and Kentucky Warbler, which forage on the forest floor. In both cases, although foraging heights were comparable, foraging strategies were found to be quite different. The Magnolia Warbler forages mostly as a gleaner, picking invertebrates off from the surfaces of leaves, while the Wilson's Warbler forages mostly as a flycatcher, making short sallies after flying insects. Ovenbirds walk along on the forest floor picking invertebrates from the ground and duff, while Kentucky Warblers hop, reaching and jumping upward to glean invertebrates from overhanging leaves. Many hours of observation revealed no aggressive interaction between members of these species pairs.

The results of our Texas work were even more striking with regard to the ecological counterpart question. Although rainforest is the iconic tropical community in terms of the diversity and specialization of its indigenous species, it is only one of many in which migratory birds pass the temperate-zone winter. Two such communities occurred at our Texas study sites: riparian forest bordering creeks and rivers and mesquite chaparral in the neighboring uplands. Our netting and banding data from Welder showed that Brown Thrashers were common winter residents, arriving in September and departing in April, holding small territories in riparian forest habitat. A close relative of this species is the Long-billed Thrasher, which is an indigenous, year-round resident at Welder in mesquite chaparral thorn forest, where it also defends small territories. If the ecological counterpart theory were valid, one would predict that the resident Long-billed Thrasher would prevent the migrant Brown Thrasher from occupying a stable niche in a primary habitat like riparian forest throughout the wintering period, forcing the migrant species to be a nomadic wanderer. Similarly, the resident Bewick's Wren at Welder failed to stop migrant House Wrens from holding winter territories (defended vigorously from other House Wrens).

The presence of a community of migrants passing the wintering period on small territories in tropical rainforest rendered key predictions of the northern home theory moot. An alternative hypothesis that better explained our observations (and those of many other tropical field biologists) was needed, and so I formulated one for the concluding section of my thesis, which stated, "Most Nearctic, long-distance migrants evolved from historically tropical species, which were able to exploit niches common to many different forest types to travel northward to temperate environments during the summer period to take advantage of lower competition for food and mates to produce more offspring in these environments than would have been possible in their tropical homes, returning southward once breeding has been completed."

This new theory shifts the value of migration to increased reproductive success as the driving force behind seasonal movement rather than overwinter survival, which is the assumed driver behind migratory movement for the northern home theory. I was not the first investigator to put forward this idea; indeed, several had suggested similar formulations, collectively referred to as "southern home" theories on the origin of migration. The important difference between my iteration and theirs was that I had all of their published data plus my own two years of intensive investigation during different phases of the migrant life cycle. The strength of the theory was that it explained how migrants could assume membership in tropical communities as soon as they arrived from temperate breeding grounds. The chief weakness was that it is much easier to understand how bad weather can force a bird southward than it is to see how potential increases in reproductive success can push a bird northward.

I spent the next forty years of my career on migration studies, attempting to understand how migration evolved and how it works. The results of these investigations and deliberations constitute the remainder of this volume.

HIATUS

Crying in the Wilderness

The Rappole and Warner paper proves
the axiom that the biggest is not best . . .
while potentially of major importance,
[it] is sadly lacking in credibility.

—R. C. WHITMORE, "REVIEW OF *MIGRANT
BIRDS IN THE NEOTROPICS*"

On the completion of my thesis and its acceptance by my committee, I was riding pretty high when I got my degree in June 1976. I felt that the data and ideas presented in it were groundbreaking and that with two out of the three chapters already in print, and the third submitted to a major journal, I should have little trouble in landing a good research/ teaching job at a major university. Indeed, I assumed that the work was of such quality that I would be able to get a paper in *Science* (the leading journal of scientific discovery in the world) and a major grant from the NSF to continue my research in the tropics.

Not so fast, my friend.

Unfortunately, things did not work out quite as well as I had expected. The third chapter of the thesis was rejected for publication out of hand by Jim Karr, associate editor of the journal *Ecological Monographs*. Karr was a MacArthur protégé who had done work on migrants in Panama; indeed, he was perhaps the acknowledged leading expert on the winter ecology of the group among new-world ornithologists. He had recently published a definitive review on migrant ecology and evolution in the prestigious bird journal the *Wilson Bulletin*, in which he confirmed the MacArthur theory of migrants as northern home birds who lived as wandering nomads during the winter season in the tropics.

The paper that I sent to *Science* presenting a southern home theory for the evolution of migration was also rejected. The editor felt that it might be OK in a bird journal but wasn't sufficiently cutting-edge for his outlet. Gene Morton, my Smithsonian guru, and I coauthored an NSF proposal to continue migration studies in Panama using radio-tracking in addition to netting, banding, and observation. The proposal received good reviews except for that written by Karr, who judged it as "poor," which killed us.

The smart thing to do when papers or funding applications are rejected is to read the reviews, rewrite, and resubmit immediately. The problem for me was that in addition to discipline and patience, a great deal of time would be required to follow that path. Resubmission to a major funding source or journal usually requires a year or more for results (at least in 1976), which, of course, are not guaranteed. Also, my family had already spent four years in nomadic movement across the continent. Asking them to go into a holding pattern for another year or two while awaiting the big break seemed more than a little unfair, especially considering that our four years of graduate school madness had been preceded by three years of military mayhem.

Given these realities, I was determined to find a job as soon as I had my degree in hand. Accordingly, in addition to the usual positions in my area of expertise, such as assistant professorships at major universities advertised in *Science* (which normally had two or three hundred qualified applicants from across the country), I applied to some positions well outside my field, like bird curator at Chicago's Lincoln Park Zoo and ecology teacher in Guam or Beirut (during the civil war).

It was at this point that John Tester, a professor in the university's Wildlife Department, called and said he had heard I needed a job. I admitted that this was true. He said he was on the board of Breck, a private high school in Minneapolis, and it turned out that they needed a biology teacher for the coming school year. He was apologetic; the salary was $10,000 for a nine-month appointment, but it was work.

I thought that given my lack of success in the international job market, I might as well see if I could handle a local one. Accordingly, I interviewed with the headmaster, John Littleford, and was offered the position, not without some misgivings on his part, I think: a PhD is a nice credential for a research scientist, less so for a high school biology teacher. I had had two semesters of teaching experience with college undergraduates, and none, of course, with younger people. Nevertheless, he took a chance on me—not least, I suppose, because of a board member's support—and I started teaching at Breck in early September with six contact hours a day: three sessions of Introductory Biology, two sessions of Human Anatomy and Physiology (for those kids already pointing toward medical school), and one session of Advance Placement Biology.

I had been introduced to the faculty by my immediate boss, Kathryn Harper, the upper school principal, some days before my initial trial by fire, but I was surprised when the world history teacher, a bearded, red-haired communist expert on Stalingrad, accosted me at the end of sixth period on my first day of classes to invite me to a meeting at his house to take place at five that evening. He didn't tell me what it was about, so it wasn't until I got there that I learned that a walk-out was the topic. Fortunately, after some heated discussion regarding the viciousness of the management (Littleford), they decided against it for the time being without bringing the matter to a vote. This action saved me from public exposure as a scab within eight hours of my arrival on the job. Solidarity be damned; there was no way I was going to go out on strike.

So our lives as members of the petit bourgeoisie commenced. We bought a house on Princeton Avenue in Saint Paul, near the University of St. Thomas. Bonnie went back to work as a registered nurse at a Catholic nursing home run by nuns of the Little Sisters of the Poor, and the kids enrolled in the Breck primary school—first grade for Jay, third for Brigetta. Soccer, ballet, piano, softball, church, book clubs, and so on became the norm.

I did not hate teaching high school. My fellow faculty were good, hardworking professionals and the students were fine, but my career as a research scientist was almost completely stalled. I continued applying for real PhD jobs, receiving enough rejection letters to paper my office walls. But after two years at Breck (rising to chair of the upper school science department at a twelve-month salary of $16,400), I knew that I had to go back to working full time on publishing and grant writing to have any chance of landing a job in the profession for which I was trained. Of course, we could not afford to do this without help, so I approached my mom for financial assistance, which she agreed to provide. Then I went to Warner and asked him to take me back at the Bell Museum as an unpaid postdoctoral fellow. Such a position, though without funding, would allow me museum office space; phone, library, and computer center privileges; and most importantly, the legitimate academic status necessary for scholarly publishing and competing for jobs and grants in my field.

The possibility of fieldwork was closed to me at Breck, but I did have the opportunity to continue the discussion through continued efforts at publication and by attending and presenting papers at national ornithological meetings. The most important of these occurred in October 1977, when I had the chance to test my "southern home" theory in a debate with the top minds in my field in the country, including Steve Fretwell, John Terborgh, Dan Janzen, and Jim Karr, among several others.

The occasion was a symposium on migratory birds in the Neotropics organized by my Smithsonian mentor, Gene Morton, and Allen Keast, a professor at Canada's Queens University. The meeting took place at the Smithsonian's Conservation and Research Center, a gorgeous research facility located in the Blue Ridge outside Front Royal, Virginia. It was at this venue that I presented the third chapter of my thesis, the one rejected by Jim Karr for *Ecological Monographs*. Discussion was lively, and although the consensus among participants likely remained the same as the earlier Smithsonian conclave of 1966—namely, that migrants were northern home interlopers in the tropics— several of the participants were at least open to the possibility of a southern home origin of migrants.

When the thirty-nine papers presented at the conference were published, a review of the volume in the *Wilson Bulletin* by R. C. Whitmore, a professor at West Virginia University, singled out our paper for special opprobrium—which, he noted, "stands out like a sore thumb" (that is, not in a positive way)—and devoted nearly half his piece to its excoriation. The gist of this diatribe was that he found our observations and tests of territoriality and short- and long-term site fidelity among migrants of the Tuxtla rainforest community unconvincing, and since these data formed the core of our argument of migrant southern home origin, he dismissed it.

The Whitmore review is worth some further consideration, as I believe it to be indicative of broader issues in the scientific community concerning the empirical validity of behavioral observations. As I have mentioned, such data often are considered mere anecdotes, and there is more than a grain of truth in this view. A brief canvassing of the behavioral literature will yield numerous examples of elaborate theoretical air castles constructed from a few field observations. Nevertheless, behavioral information can be critically important, as it tells us how the organism is interacting with its environment in real time, which can, in turn, lead to new insights into and a deeper understanding of fundamental aspects of ecology and evolution. In the case of migrants, the discovery of seasonal territoriality in several species wintering in tropical rainforest potentially represented just such a watershed moment. The northern home theory of migrant evolution did not allow for this kind of behavior. If true, long-term territoriality by male and female migrants in stable tropical habitats required a new theory. Which means that the importance of our paper in the Smithsonian symposium came down to whether we had successfully made the case that several migrants in the Tuxtla rainforest community were, in fact, territorial.

The data presented by us included the following types: within-season and between-season capture, recapture, and resighting of individuals on the same small plot of ground within the rainforest; use of vocalization, stylized displays, and attack in apparent defense against trespass by both free-flying birds and staged intruders (caged birds, mounts, mirrors); regular advertisement by means of vocalization, including song, by both male and female territory owners; observation of nonterritorial birds (floaters) as occasional invaders throughout the winter season, which would take over a territory if the owner were removed; mapping of the territory boundaries; and use of playback to stimulate response by territory owners.

It is quite true that we were able to collect only a few of these kinds of data for some species, like the Yellow-bellied Flycatcher and White-eyed Vireo, while for others, like the Hooded Warbler and the Wood Thrush, we had seen nearly all of these behaviors, thereby confirming, in my view, long-term, interseasonal territoriality for perhaps four or five species while providing data strongly indicative of this behavior for another fifteen or sixteen.

Thus, we, at least, felt that we had presented enough to allow us to seriously question northern home orthodoxy. Whitmore begged to differ. He claimed that we had not confirmed territoriality, placing the term in quotation marks throughout his review as though it were an obviously false assertion.

The lesson I learned from this dismal encounter was twofold. First, for whatever reason, some people will not be convinced by any data or argument that you can present. You are wasting your time if you try. Second, if their critique is made in some public forum, such as a meeting or in a publication, you must attempt immediately to present a rebuttal in an equally public manner, verbally and/or in writing. Most people in the audience will have no set opinion, and you must make your case to them. In essence, they represent the jury. With regard to Whitmore's review, I submitted a rebuttal to the editor of the *Bulletin*, but he refused to publish it, so I was forced to argue the issue in a number of subsequent publications.

Rejection is painful, and the old saw "If you can't stand the heat, get out of the kitchen" is small comfort, especially when you aren't even in the pantry. Nevertheless, the best action to take when in this position is to consider the weaknesses of your argument and attempt to address them through continued experimentation and discussion in a scientific forum. Having returned to academia in the summer of 1978, I now had the opportunity to pursue this avenue for at least the coming year. One way or another, I would have to have a job by mid-1979 or go back to teaching high school and give up my hopes of a research career. I went to work accordingly, writing and submitting papers to several journals, and applying for every research or college teaching position I could find.

By early 1979, with no job offers, I was beginning to feel intense pressure to find some position, any position, when I got a call from the chairman of the science department of LeMoyne–Owen College, a school mostly for Black students in Memphis, inviting me to come down and talk about their opening for an assistant professor in biology.

Memphis that February was not lovely as I flew in from Minnesota. From the air, it seemed mostly to be mud brown. Nevertheless, I was excited to be at my first college-level interview, and it went well. I met the chairman, an affable Pakistani, and the other faculty, who seemed happy to see me, since I represented some relief from their crushing classroom loads. I gave a seminar for the students, whom I found delightful. At supper at his home, the chairman offered me the position, which I tentatively accepted, asking what the salary might be. He said it would be $13,000 on a nine-month appointment. At that, I balked. I said there was no way I could move my family to Memphis for less than $14,000. He said he was sorry, but if he were to offer me that, it would upset his other faculty, since that was what they were making.

The next day, I flew back to Saint Paul. The chairman said he would think about our discussion and call me in a day or two. As it turned out, he didn't budge, and fortunately, neither did I. In the meantime, I called my sister Francesca's husband, Bryan Miller, to get his opinion. Bryan was chair of the biochemistry department at the University of California at Davis, an NSF panelist and high-end research scientist. When he heard my tale, he said I was insane to consider taking a job at a small college. "They will work you to death doing classes, labs, and administrative duties with no time for research," he warned. "You would have been better off staying at Breck than taking such a position. Just keep your courage and hold on. Something will come through."

And it did—through an odd piece of serendipity. Arguably, the best research job in 1979 for a newly minted PhD in ornithology was a postdoctoral position at the University of Georgia's Museum of Natural History, which had been advertised in a November 1978 issue of *Science*. I knew no one there and figured I would have little chance among the likely pool of a hundred or so well-qualified applicants, but I dutifully put together my application and contacted my references for supporting letters anyway. I was stunned, then, to get a call from the director of the museum, Josh Laerm, in March inviting me down to Athens for an interview and even more floored to be offered the post a month or so later.

Josh subsequently told me that my selection was due largely to his graduate student, Joe Meyers. Josh was a mammalogist by training with a PhD from the University of Illinois. The bird postdoc was one of four positions that he had available through a grant from the state to evaluate Georgia's endangered species list, the others being for a herpetologist, an ichthyologist, and a mammalogist. Since he didn't know much about birds, he had Joe, who was doing a PhD on avifauna of Okefenokee Swamp, look over the ornithologist position applications. As it happened, Joe had attended a paper I gave at the 1978 AOU meetings and had come up after the session was over to discuss some of my findings in more detail. Based on this brief exchange, he picked me out of the application pile as a person with whom Josh should at least talk. Ironically, R. C. Whitmore, author of the devastating review of our Smithsonian symposium paper, was also an applicant for Josh's opening.

A further break of major proportions occurred shortly after my invitation from Georgia. Gene Morton called to offer me the chance to prepare a literature review on the topic of migratory birds in the Neotropics, which would ultimately be published as a book. To prepare the review, funding for two years was provided by grants from the World Wildlife Fund (WWF) and the USFWS. I called my new boss to find out if it would be OK with him if I did my Georgia postdoc at the same time as I was doing the literature review, and Josh said he thought that would be fine. As a result, in the spring of 1979, I was finally able to begin my career as a research scientist with a specialty in migratory birds.

CHAPTER 14

MRS. FISK

I cannot rest from travel: I will drink
Life to the lees: all times I have enjoy'd
Greatly, have suffr'd greatly, both
with those
That loved me, and alone

—ALFRED LORD TENNYSON, *ULYSSES*

The bulk of my time in Georgia, 1979–81, was spent surveying the state's avifauna to determine which birds should be considered for inclusion on the list of threatened or endangered species. This activity was not directly relevant to my study of the biology of long-distance migrants for the most part, but I was allowed a great deal of freedom by my boss, Josh Laerm, director of the University of Georgia's Museum of Natural History, not only to prepare the Nearctic migrant literature review but to continue my research on migrants in Veracruz if I could find the funding.

The chief enigma of migrant winter ecology in the tropics is the behavioral dichotomy between migrants in open habitats and marginal second-growth scrub, where they often occur in seemingly itinerant flocks in contrast with the same kinds of birds in forest, where members of at least some species show long-term site occupancy and territoriality. As I have noted in earlier chapters, many students of tropical ornithology did not even realize that migrants occurred in forest.

This notion was quite evident in the first Smithsonian volume devoted to migrant wintering ecology and conservation, wherein most of the elite group of tropical ornithologists who convened for the symposium in April 1966 were obviously somewhat dismissive and nonplussed by the topic of migrants in the Neotropics anyhow, an idea not as surprising as it may at first seem.

Consider that, just as Warner and I found when we were looking for sites back in 1973, forest in much of Mexico and elsewhere in Middle America and the Caribbean was simply gone. In many regions, the only remaining bits were strips along rivers or steep mountain slopes. The only reason there was any left in the Tuxtlas had to do with mountainous terrain and heavy rainfall, the combination of which made some parts difficult to access.

Nevertheless, when we first started working there in the early 1970s, about 70 percent of the forest had been cut; by the 1990s, 90 percent of the forest was gone, and most of what remained was at higher elevations or inside the vast craters of the Santa Marta and San Martín volcanoes. In addition to a lack of forest, there is also the fact that most ornithologists do not go to the tropics to study birds from the temperate zone; they go there to investigate the spectacular tropical avifauna, and since migrants often are not their main focus, they devote no special attention to them. This approach guarantees that the majority of their migrant sightings according to their field notes will be in open areas and marginal second growth, where many migrants are common, easy to see, and often constitute a prominent portion of the bird community.

In a preceding chapter, I criticized Whitmore's review of our paper in the Keast and Morton volume on migrants in the Neotropics for devaluing our reports on migrant use of forest. However, he did point out, correctly, two important questions regarding songbird migrant winter ecology—namely, why most songbird migrants are seen in marginal habitats in winter and whether there is any evidence that migrants that live in marginal habitats have lower fitness (that is, produce fewer offspring) than birds of the same species wintering in primary habitats like rainforest. These questions are critically important to our understanding of migrant ecology and evolution. They are also difficult to answer given the standard array of field biology techniques available—expensive, time-consuming, and not easily submitted to normal statistical analysis. Despite these formidable conditions, Mario Ramos and I, along with a splendid team of colleagues and students, set out to try to address them when I got myself situated in a research position that allowed time to prepare and submit some grant proposals.

The first such grant came through for us in September 1980 by way of Tom Lovejoy, then vice president of science at the WWF. A former MacArthur postdoc and graduate student colleague of Gene Morton's at Yale, Tom had a strong interest in tropical conservation in general and migratory bird conservation in particular, and he provided some WWF money for Mario and me to get back out in the field in the Tuxtlas.

The plan was to go back to my study area at Playa Escondida in southern Veracruz and mist-net at my old sites to see if we could find any of my banded birds from our 1973–75 work and to compare capture-recapture/resighting rates for members of the same wintering migrant species in forest and neighboring marginal habitats. In other words, we were trying to get an answer to the question posed by Whitmore: Is there a difference in the persistence of migrants in marginal habitats when compared with forest habitats? We already knew that members of several migrant species in forest remained on small territories throughout the season, but we had not sampled enough in marginal scrub and second growth to get a good comparison.

Since our time together in the Tuxtlas, Mario and I had both landed university jobs and become major advisors to a few graduate students; Mario in particular had already secured a reputation for excellence on the world stage despite his youth. In 1980, he was only thirty-one and already president of the Neotropical Ornithological Society (NOS), a member of the International Ornithological Council, and senior research ornithologist for the elite Instituto Nacional de Investigaciones sobre Recursos Bióticos (INIREB).

We decided that we would head for Playa from our respective home institutions in late November and plan to stay until late December. About a month before our planned departure, I got a call from Tom Lovejoy, who wanted a favor. He said that he would really appreciate it if a friend of his, Erma "Jonnie" Fisk, could join us to help out with our fieldwork. I said, "Gosh, Tom, I don't know. How old is this lady?" He said he wasn't sure—maybe early sixties. I said that working in the Tuxtlas could sometimes be pretty rough: isolation, primitive living conditions, rain, mud, snakes, and so on. He said it was no problem. She had run a bird-netting field study all by herself for months in the Chiricahua Mountains of southern Arizona and had also been on team banding forays to Belize. Well, what was I going to say? His organization was providing our funding, and if he thought she could do it, then who was I to question him? So I told him I would give her a call and invite her to come along.

After hanging up, I thought I remembered who she was and looked her up in the bibliography I was building for my Nearctic migrant book. Sure enough, she had published a paper on Least Tern nesting sites on flat gravel rooftops near her winter home in Homestead, Florida. Anyhow, I called her up, introduced myself, and asked her if she wanted to come down to the Tuxtlas. She said sure. I described the living and working conditions there, which she said would be no problem, and we agreed that I would pick her up at the Veracruz airport at a date and time she would provide me later.

On November 12, I set out from Athens on the two-thousand-mile drive to the Tuxtlas for a month of fieldwork. I should mention that in so doing, I was abandoning Bonnie, who had to not only keep Brigetta (almost twelve) and Jay (nine) on track but care for our four-month-old son, Nate, who had been born five weeks prematurely back in July. She was also working as a private nurse for our next-door neighbor who was dying of cancer. Managing the home front during my many prolonged absences over forty years of fieldwork was never easy.

Accompanying me on the trip were two of my graduate students, Kevin Ballard and Lisa Neuwirth Yoder, along with Lisa's husband, Mike. I had rented a brand-new (twelve miles on the odometer) 1980 Crown Victoria from a local Athens dealership at an excellent price. The Crown Vic is probably not the vehicle of choice for Mexican mud tracks, but it has plenty of power and storage space (especially with our added car-top carrier) and is a great ride. I also took out bumper-to-bumper insurance.

The trip was relatively uneventful, and we arrived at Playa at about four in the afternoon on November 15. Our friends Raul and Julieta Garcia and their factota, Angel Toto and his family, were all well, and the patio and living units appeared much the same. Not so for the forest. Nearly five years had elapsed since I was last there, and I was stunned to see the changes that had occurred to my main study site.

What had been twelve acres of undisturbed rainforest in March 1975 was now mostly a hodgepodge of pasture, cornfield, and scrubby second growth with an island of about four acres of forest left. This discovery was personally painful, resulting in a couple of pages of florid ruminations in my notes, but gave us an opportunity to do a "before and after" study of migrant species and territories that had been on the site when it was entirely forest, 1973–75, as compared with the mixture of marginal second-growth habitats and forest in 1980. It also made me realize that if I wanted to do any long-term studies of migrants in tropical forest, I would have to find another site, one that was protected by either law or circumstance.

The most obvious possibility was UNAM's biology station, where Mario and Oehlenschlager had done most of their work in the early 1970s. Accordingly, I headed over there on the morning of the sixteenth, and I found that the station too had changed. UNAM's Instituto de Biología had put quite a bit of money into the place since we had worked there. Thanks to the efforts of then director Dr. Arturo Gomez-Pompa, it now boasted labs, student dorms, offices, and living quarters for the resident director, Dr. Alejandro Estrada, a young up-and-coming member of the Instituto faculty with a PhD in behavioral ecology from Rutgers, where he had studied howler monkeys for his thesis research.

Dr. Estrada was decidedly cool throughout our introductory interview and not inclined to allow us to work at the station, antipathy that stemmed apparently from the failure on the part of our Minnesota research group to deliver on promised specimens and catalog copies. I promised to rectify this to the best of my ability, at the very least providing him with complete duplicates of all data collected previously as well as anything further that we might accumulate. During additional discussion, I made the case that whatever our collaborative failings had been, actual or perceived, the chance to do follow-up work on birds of the station's forest, now one of the few remaining lowland blocks in the entire Tuxtla region, was invaluable. He agreed in the end, although not with good grace, and I put Lisa and her husband to work setting up a grid with a net every 150 feet on a ten-acre square in the vicinity of the station headquarters. Back at Playa, Kevin and I established a similar netting grid on our old study area as well as in neighboring pasture and scrubby second growth. We quit work at dark and headed back for supper at the restaurant patio, where, to my delight, we found Mario and six of his students just arrived from Xalapa, the capital of the state of Veracruz and home base of Mario's employer, INIREB.

Mario was already aware of the devastation stalking Tuxtla rainforest as well as the difficult relations with Dr. Estrada. He explained that a good part of the friction came from political fights for resources and influence at the highest levels of government between UNAM's Instituto and Gomez-Pompa's upstart INIREB, which was patterned conceptually on the Smithsonian—although with a much more applied aspect to its studies as opposed to pure science. Anyway, he said, it was time to see if we couldn't locate lowland forest for our research purposes somewhere else in the area.

The Tuxtla region centers on two volcanic massifs along the immediate Gulf Coast about sixty miles south of Veracruz city: Volcán San Martín in the northwest and Volcán Santa Marta to the southeast. Playa and the biology station were located in the San Martín portion of the Tuxtlas, where almost all lowland forest had been cut by the early 1980s. The Santa Marta section was far more rugged and difficult to access, which was why Mario and I thought we might be able to still find some undisturbed *selva* there where we could continue long-term studies of Nearctic migrants and tropical residents living in rainforest.

We left at six the next morning (November 17) to do some reconnaissance in the remote Santa Martas, traveling south to Catemaco, then east along the shore of Lake Catemaco to the old soft drink factory at Coyame, and from thence up into the highlands (above three thousand feet) to the small village of Bastonal. There we left the truck and proceeded on foot across pasture for a couple of miles to the edge of a steep ravine, from which we could see apparently untouched forest lining the gorge and broader valley below. Following a hunter's path, we worked our way down several hundred feet to the river, a cataract really. The forest was indeed untouched, although it was not *selva*. The vegetation was more typical of highland evergreen forest—cloud forest, actually—including tree ferns, sweet gum, and oak, and although several migrant species were present, Hooded Warbler, Wood Thrush, and some others typical of lowland rainforest were not.

Mario noted that a park service ranger leading an exploratory group into the valley the previous year had died of a heart attack and had to be lugged out by his fellows.

On the steep climb back out of the valley, a *norte* hit, with high winds and downpours, drenching us. Although there were some periods when it let up a bit, the rain never completely stopped so far as I was aware for the next ten days.

Mario and four of his students headed back to Xalapa on the morning of the eighteenth, leaving me with two to help for the next few weeks: Sergio Barrios Monterde and Jorge Vega Rivera. These young men became graduate students of mine in the years to come and close and deeply valued friends to the present day.

Mrs. Fisk was due to land at the Veracruz airport at 7:30 on the evening of November 20. The original plan was for her to fly down with Gene Morton to join our work party at Playa, but the mother of Gene's wife (at the time), Letty, was taken ill, so Gene had had to cancel, which I had found out from Bonnie when I called her from Welder on the thirteenth. This circumstance left Mrs. Fisk to make the long trip from Boston to Veracruz on her own, to be picked up by a man whom she had never met—me. At least I assumed that that remained the plan.

I got up at five on the morning of the twentieth and worked with Jorge setting nets in scrubby pasture in the rain until noon. Then we set out for Veracruz to put Jorge on the bus for Xalapa, after which I would head over to the airport. I arrived in plenty of time because I liked the margaritas and food at the airport restaurant, which had a buffet of enchiladas, refried beans, guacamole, and so forth.

The flight was coming from Mexico City, on time at 7:30 p.m., and I was standing in the receiving area as the passengers walked in from the plane, which had unloaded on the open runway. None looked like a late-middle-aged American lady traveling alone. I went to the Mexicana Airlines desk and asked them to check the passenger list, which they did. There was no one by the name of Mrs. Erma Fisk. I was not too surprised by this. Snafus are common in foreign travel, but I thought I had better see if I could find out for sure what was going on. In the early 1980s, telephone service to the States was not routine. There were no phones of any kind in the Tuxtlas outside San Andrés, Catemaco, and Sontecomapan, and even in these metropolises, you had to go to a special office where you got in line to have an operator put through an international call.

However, there was a phone where you could make international calls, collect, at the airport, and I managed to get through to Bonnie. She said that all systems were go so far as she was aware.

So what to do? Back then, the Veracruz airport was not a busy place. One more flight came in from Mexico City at 9:30 that night, but Mrs. Fisk was not on it or on the passenger manifest. Since it was the last flight due in for the day, the whole place shut down and locked up within a half hour of its arrival. By the time I went out the doors, the only persons I could see in the lobby were cleaning staff.

It's a little tricky to understand the choreography of this next bit, so pay attention. The airport was located about a quarter of a mile off the main Veracruz-Xalapa highway. As you approached on the road to the terminal, you were given a choice of lanes—an inner lane for parking and an outer lane for picking up arrivals. As you departed parking, you were again given a choice of lanes, an outer one that was the exit or an inner one that joined the incoming lane from the Xalapa-Veracruz road, which then led back to the two-lane choice of arrivals or parking. I was tired or maybe not so used to margaritas, I guess, but for whatever reason, I missed the merge right for the airport exit and came back around to the two-lane choice of parking or arrivals. Cursing my stupidity, I took the arrivals turn and started to pull on by the front of the airport. As I did, I noticed a trim lady in slacks, a red sweater, and a jaunty tam standing alone beside an enormous suitcase in front of the large glass doors of a now completely deserted airport (and empty parking lot). I stopped the truck, got out, and approached, saying, "Mrs. Fisk?"

Jonnie later admitted to me that this was perhaps the most dispiriting moment of that memorable trip. Let's just get it right out there: at five foot seven and 145 pounds soaking wet (not unusual during much of my years of fieldwork), I was not physically prepossessing at my best, and after six hours of netting in the rain, a four-hour drive to Veracruz, and four hours of hanging around the airport, I was nowhere near my best. Mrs. Fisk, of course, knew nothing about my day, and she had a different image of how a college professor in mufti should appear. As she later explained, she imagined that Dr. Rappole would be a tall, well-groomed, distinguished-looking individual with (maybe) a nicely trimmed goatee. Not in tweeds, of course, because of the tropical heat, but perhaps a nice pair of khaki slacks, guayabera, and leather sandals. Certainly, a friend of Tom Lovejoy's would be expected to maintain some standards, even when working in the field. She was not expecting the shrunken, scruffy specimen in flip-flops, muddy Levis, and a formerly white T-shirt who approached out of the rainy dark from what appeared to be some sort of salvaged vehicle. Of course, she didn't put it that way. She was, after all, an upper-crust Vassar girl of the first water.

Despite all of this and her own exhaustion, gathering her (considerable) moxie, she responded in the affirmative. "Great," I said. "Glad to meet you. I'm John Rappole. Sorry about the confusion, but happy that you are here." I threw her suitcase into the back of the truck, and we took off for the city to look for hotel rooms. En route, she explained what had happened. She arrived early in Mexico City and was put on the 4:30 p.m. flight to Veracruz instead of the 7:30 one. On arrival, and up until she was kicked out of the airport, she looked for distinguished gentlemen, but observing none had assumed that something had happened.

"What would you have done had I not appeared?" I asked.

"Oh, I would have taken a taxi into the city, spent the night, and then flown back to Cape Cod," she said—pretty coolly, I thought. There weren't going to be any taxis or traffic of any sort there for about ten long, dark, wet, cold hours. That was Jonnie, though. After her husband, Brad, had died, she figured not much worse could happen to her and didn't really care. She was an amazing woman (passing away in 1990) with awesome depths of zest and courage, of which this little episode hardly scratched the surface, as I was shortly to learn.

We stopped at a couple of hotels in the city and could find no rooms at that late hour, so we set out on the long drive to Playa, pulling in at about 2:30 in the morning of the twenty-first. I put Jonnie into Lisa and Mike's room (two double beds) for the night, and we all settled down. As Warner was wont to say, "Sleep fast, boys; it's daylight in the swamp."

"For the rain it raineth every day." Shakespeare had it right. Although he knew nothing about the Tuxtlas, he evidently knew something about the effects of a few rainy days on one's morale. The twenty-first was the fifth straight day of almost continuous rain, and though we (thankfully) didn't know it, five more days of it lay ahead; indeed, it rained on twenty of our twenty-five days in the Tuxtlas.

Since both Tom Lovejoy and Mrs. Fisk herself had told me that she was an experienced bander, I had decided to keep her out of the mud and rain and give her Bonnie's job as recorder, making sure that all of the data were collected and properly entered for each bird captured. She would have none of it. "I didn't travel thousands of miles to sit in a chair!" she said. However, there was no way she could negotiate the slippery slopes and gullies at Playa, so I took her over to the station, where it wasn't so tough, to help out Lisa and Mike with their nets.

Figure 14-1. Professor Rappole in full field regalia. Cutting mist net lanes through spiny second growth was hard on rain gear, forcing me into various makeshift substitutes for the tattered remnants of my original garb. On seeing this photo, Bon said that while she got the body bag, there was no possible excuse for the ridiculous head covering.

Things were not going well there. Dr. Estrada may have given us permission to work, but he was determined not to make it easy. He had guards following Lisa and Mike everywhere they went, which made it especially problematic for Lisa to find a private spot to relieve herself over the course of her day in the field. Also, he wouldn't give them a key to the lab so that they could do their banding and recording in a covered place. They were glad to have Mrs. Fisk's help, but she soon found that even the relatively benign terrain of the station was too much for her after a week of rain. In her defense, I should say that I had forgotten to include chocho palms in my description of likely problems for field researchers in the Tuxtlas during our pretrip telephone conversation.

The chocho (*Astrocarium mexicanum*) is a ubiquitous part of the *selva* understory. It grows as a single bole a couple of inches in diameter with fronds at the top to a height of ten or fifteen feet. The trunk is densely covered with brittle spines one or two inches in length. Slipping and sliding along muddy arroyos, one's first reaction is to reach out for balance and "shake hands with *el chocho*," resulting in seven or eight thorns broken off a half inch or more in one's flesh. Every long-term researcher in the Tuxtlas will carry chocho spines buried deep in various parts of their bodies to the grave. I

Figure 14-2. Mike Yoder and Jonnie Fisk outside her Playa room on November 26, 1980.

myself have two in my right wrist and one in my right thigh that became a part of me decades ago, despite the efforts of both ancient (Popoluca) and modern medical science to remove them. Mario had two in his head that we never were able to extract, and on this 1980 trip, Lisa's husband, Mike, punched one into his right arm that immobilized him for net clearing throughout the remainder of his stay.

Since she was averse to recording duties and could not help Lisa and Mike run our nets in the forest, I set a couple nets up for Jonnie in the small piece of lawn outside the lab, which she ran on her own. The one semipositive result of her work at the station was that she quickly made friends with Dr. Estrada's wife, Rosie, a Canadian by birth who was delighted to have a woman of Mrs. Fisk's intelligence and knowledge of the world to converse with. It was not long after Mrs. Fisk's arrival that the previously denied keys to the lab were provided to our crew. There, Rosie and Mrs. Fisk would sit sipping coffee for hours and discussing topics of interest, too much of which seemed to concern my failings, professional and personal. Then in the evenings, Mrs. Fisk would sit with me on the restaurant veranda and give me the benefit of their mutual deliberations coupled with her years of diplomatic experience. In her view, I was something of a mulish little toad (my interpretation), running around giving orders to everybody without considering their situation or point of view. She tried hard to make me see how some slight modifications to my demeanor and behavior might make things better for all concerned. I'm sure she was right. I had little time for people's "issues" and still don't.

I knew well enough that Estrada's professional life was not easy. I was certain that he was getting pressure from above to extract what he could from the gringo scientists while still somehow letting them do the sort of work, which, when published, would promote the exceptional value of UNAM's station to the world. He was a first-rate biologist and personally fearless, as refuge managers often must be. Death threats from local people trying to use the resources on your refuge are a matter of course if you take protection seriously, as Estrada did. A story that summarizes his personal bravery tells of how he and his wife and kids were stopped by a roadblock (log) as he was transporting the monthly wages for the staff to the station. When the chief bandito approached the car to demand the payroll, rifle at the ready, Alex threw the vehicle into reverse, backed around the corner, and K-turned toward Catemaco. Now that takes balls.

Such traits were of no use to me, however, and with his constant badgering and surveillance, he made our work difficult and the lives of Mike and Lisa miserable. Over a beer, Alex and I might commiserate concerning our respective problems, but the bottom line for me was that Mario and I had been given one of the few grants available to do fieldwork on migratory birds in the tropics; a considerable amount of planning and personal commitment by several professionals and students had gone into the preparation. Our time for gathering the data needed to answer the questions posed was limited, and the execrable weather wasn't helping. Tlaloc (rain god) reigns often in the Tuxtlas, and this season was no exception.

The rain and work continued, and most of our crew were sick at one point or another—usually intestinal disorders, although I had the flu for a day or two. On the twenty-eighth, it rained off and on during the day but started to clear a bit later in the afternoon. Kevin and I were running nets near Playa when Mrs. Fisk caught me at about five in the evening and said she would like to take advantage of the break in the weather to do a little bird-watching down the mudslide that had been our road. I wished her luck, and we headed out to run a last check and close up for the day.

At about six as I was headed back to my room, Kevin came up and asked, "Where's Mrs. Fisk?"

Whoa! Good question. I was mortified to say that she had gone down the road birding about an hour ago and I had forgotten all about her. I ran to my room to get my flashlight, and we headed off in the dark to see if we could find her. We hadn't gotten far when, to my immense relief, we saw her coming toward us. Running up to her, we were horrified to see blood on her wrist and neck and her generally disheveled state.

"My God, Mrs. Fisk! What happened?"

She gave an exhausted chuckle and said, "Well, believe it or not, an attempted rape."

When we had gotten her to her room and cleaned and bandaged her wounds—which were, fortunately, superficial—she told the following appalling story: "After leaving you, I walked down the road, birding along the way, to the turnoff for Jicacal. There I ran into a fisherman coming out of the village, headed for the bathing stream. We struck up a

Figure 14-3. View of the white sand beach and fishing "village" of Jicacal from Playa's patio with Volcán Santa Marta in the misty distance.

conversation, but as we were walking along, I noticed a man out in the pasture waving his arms and yelling at me, which made me nervous, as I thought he was telling me to get off his land, so I turned around and started to head back up the road toward Playa."

We later found out that this man was trying to warn her of her danger.

"The young fisherman went with me, and once we were around the bend and out of sight from the village and the man in the pasture, he put his arm around me and tried to kiss me, which I found quite funny for some reason. Laughing, I tried to push him away, gasping *bobo* [fool] and *abuela* [grandmother]. However, he then shoved me violently onto my back and tried to tear off my pants, which was complicated because I had on a body poncho cinched around my waist with a thick leather belt. After some fumbling and cursing, he ripped my watch off my wrist and my beautiful Leitz binoculars from around my neck [hence the bloody scratches], kicked me, and ran off down the road."

Overwhelmed with shock, dismay, and guilt, I apologized abjectly for my stupidity and lack of oversight, recognizing full well that I had betrayed my responsibility for her. I then told her that, provided she was feeling OK, I would leave to talk with Angel about what he might suggest we do about the crime and stolen goods.

After hearing the story, Angel said that he would go down to Jicacal and see what he could find out. Feeling useless and ashamed, I went back to meet with the crew at the restaurant and wait for his return. He came back three hours or so later and reported that he had talked with people at the fishing village, none of whom seemed to know anything. Then he had walked the few miles to the nearest bar (a thatched hut at the village of La Palma) and talked with some more people there, again with no luck. He had no suggestions regarding what to do next.

The next morning, I got up at about six, got the crew working to set nets in forest and second growth, and started running a visual census in pasture and scrub along the road past Jicacal toward the station. Walking along, however, I could not get the enormity of what had happened out of my head or how useless and ineffectual I felt. "There really isn't much we can probably do," Angel had said. "No one is going to inform, so we might just as well be thankful that nothing worse happened and let it go." Good advice, I supposed. The nearest police were thirty miles away in Catemaco, and I knew from experience that the investigation of crimes committed in the bush was not considered a part of their brief. Given what evidence we had, they would not even bother to make a report, let alone drive out to Playa. The best plan was to count our blessings and focus on the few remaining workdays left to us.

Walking along by yourself watching for birds, I have found, is not good for your peace of mind if you have a worrying problem. After a couple of hours of this broken-record kind of nonthinking, I decided that doing nothing was bullshit. First there was my responsibility to Mrs. Fisk. She was my guest, and I had allowed a terrible thing to happen to her. Doing nothing would just compound the offense. Then there was the fact that we now had a known thief and potential rapist in our midst, unapprehended and unpunished. Mrs. Fisk was not the only woman who spent time alone in the woods. My student, Lisa, did as well of course, as did Mario's student, petit Isabel Carmona. And then there were Angel's daughters, Gloria, nineteen; Maria, sixteen; and Virginia, four-teen. They would be unlikely to be so foolish as to wander around alone, but still. Am I supposed to just say "Fine, girls, take your chances"? Obviously, I couldn't do that. But what to do? Eventually, I just thought, "What the hell. She saw her attacker, and it's likely that he's still right down the road, either in his hut or out in one of the fishing boats. Penniless, friendless, and out in the middle of nowhere, it isn't like he can just leave. There isn't even any bus service within a day's walk. So if Mrs. Fisk is amenable, let's see if we can find him." With this resolution, I walked back up to Playa, found Mrs. Fisk and Angel, and asked them to go with me down to Jicacal. They agreed.

Here I need to pause the story for a moment to give you some information on Ejido Jicacal. As mentioned in chapter 3, an *ejido* is a sort of commune established on govern-ment land in which each *ejidotario* (male householder) has the right to use a portion of the *ejido's* property, usually a few acres available for raising corn, beans, and chiles at little more than subsistence level. Ejido Jicacal was a little different in that in addition to farming, the members also fished, for which purpose the *ejido* owned several large open boats. Fishing, however, was seasonal, and at these times, more people were needed to man the boats than the agricultural activities of the commune could support long term, so men, mostly homeless drifters from Veracruz city or beyond, were recruited.

This dichotomy was clearly represented in the structures housing the two segments of the village's population. Of the twenty or so huts scattered along the beach, about half housed the *ejidotario* families—that is, farmers and their wives and kids. It was one of the men from these, by the way, who had tried to warn Mrs. Fisk. The other half of the huts essentially were temporary housing, consisting of a stick frame with a roof and walls of palm fronds located just above the high tide line on the beach. Four or five men lived in each of these, each with his own hammock slung between two posts. There was no other furniture of any kind, and none was needed, since few of the men had more possessions than would fit in a shopping bag.

Walking into the *ejido*, I asked Angel to locate and introduce me to the *patrón* (head man), which he was soon able to do. We then explained what had happened to Mrs. Fisk the previous evening and asked his permission to have her examine all the men to see if she could identify her assailant. The *patrón* did not want to do this, so I asked Angel to explain that we would come back with the Catemaco police to help us do the search if necessary—an idle threat on my part, but apparently with just enough plausibility to get his compliance. So Mrs. Fisk and Angel, accompanied by the *patrón*, worked their way down the beach, entering each hut to examine its inhabitants. It was on entering the last hut that Mrs. Fisk thought she recognized her attacker: a shirtless, stocky young man in shorts with shoulder-length black hair. I asked, "Are you sure that this is the man?" She answered that she was pretty sure but that it was dusk and things had happened fast. I said, "Please be absolutely certain because this man's life is never going to be the same if you make this accusation." She agonized for a while and then decided that she was, in fact, sure. With this information, Angel walked off to talk things over with the *patrón*. In a few minutes, he returned and said that the *patrón* says this is not the guy. The kid she had identified was not a drifter but the son of one of the *ejidotarios*, definitely not the criminal. The attacker, evidently known to everyone in Jicacal, was not here right now but would be back later. I said to the *patrón*, "OK. We're going to the Catemaco police station to get a couple of constables to follow us back. When we return, you show them where the man is, and they will take him into custody."

We then left to take Mrs. Fisk back to Playa and borrow Raul's truck for the drive into Catemaco. At the police station, two fat cops in rumpled pale-blue uniforms were happy to accompany us on our little outing. Grabbing a couple of old M1s, they piled into their van, rifles sticking out the windows, and set out to follow us. When we got to within about a mile of the Jicacal turnoff, I pulled over, got out, and went back to the cop van. I said that I thought it would be best if I left them there by the side of the road while Angel and I went to find the *patrón* and see if the suspect was present. Great, they said, and we set off. We were able to find the *patrón* pretty quickly. He said that our man had just gotten off his crew shift, was back at the bathing pool in the pasture north of the village cleaning up, and should be in his hut soon. Then we went to collect the police.

Here, things started to get a little like a Max Sennett comedy, as Mrs. Fisk says in her account of these events in her book, *Parrots' Wood*. Certainly, there were some keystone cops moments, beginning when we got back to the vehicles to find them empty with no one in sight. Frantically, I started searching and shortly found that they had wandered off in search of shade; found some orange trees, the fruit of which they sampled; and then settled comfortably for a little siesta. By the time we got them rounded up, collected their firearms, and had them ambling back toward the village, a half hour or more had passed, and I was concerned that circumstances on the ground might have changed. The *patrón* met us at the southwest outskirts of the village, and said he thought things were OK. The guy had returned and was now alone in his hut, his cohabitants having been forewarned to make themselves scarce. The posse then proceeded quietly down the row, the *patrón* leading the way, while the demeanor of the cops underwent a radical change from *Car 54* to *Serpico* as they tiptoed through the sand in his wake, eventually arriving in front of the hut. The interior layout having been explained to them, the cops took up stations on each side of the rickety collection of slats that passed for a door, and with a glance and nod to each other, they burst through. Angel, the *patrón*, and I stayed outside, well clear of the action, and not wanting to be collateral damage. There were sounds of a brief struggle, and then the prisoner was roughly frog-marched through the door, his arms pinioned, his head pulled back by his hair. No Miranda rights, I guess. Dressed only in a ragged short-sleeved shirt and shorts, he appeared to me to be a little taller, older, and thinner than Mrs. Fisk's earlier suspect. I had detailed Mario's student, Sergio Barrios, back up to Playa to bring Mrs. Fisk down for a confirmatory ID once the collar had been made, and they now appeared. She took a close look at him and said she was sure that this was her attacker. Of course, she had said that about the other man as well.

I worried that the whole thing might be a setup. The *patrón* had said that they knew this guy was a bad actor. He had shown up for the season a few weeks before, on the lamb from Villa Hermosa some thought, but nobody knew anything about him really. This account sounded too pat—exactly what might be concocted to sacrifice an unknown to save a son of the community. I asked the *patrón* to accompany me into the hut and show me the man's hammock. We went in and found the interior much as I have described, a thatched roof and walls with a sand floor, completely bare except for plastic hammocks slung for five men. Hanging from a nail in the suspect's headpost was a brightly striped polyester net shopping bag. Inside I found his ID papers and two ladies' watches, one of which turned out to be Mrs. Fisk's Timex. No sign of her thousand-dollar binoculars. I stood there considering for a minute, then bent down to the base of the post, and after scrabbling around a bit, I found the binos wrapped in plastic and buried under about six inches of loose sand. Whew!

The capture took place on a Saturday. Since it was not an emergency, we were told that we should come back after the weekend for the "trial" (basically a deposition). So on Monday morning, Mrs. Fisk and I set off for Catemaco, accompanied by Sergio to translate.

Mexican law is basically Napoleonic law. A case of this kind is brought before a magistrate who interviews the victim and any witnesses and then makes a decision—no jury. We got in to see the magistrate in his office at about 10:00 a.m.—just me, Sergio, and Mrs. Fisk. Neither the defendant nor anyone representing him was present. The magistrate began by asking Mrs. Fisk for her name and age, to which she responded, "Mrs. Erma J. Fisk." Then, in a softer voice and looking directly at me, she continued, "Seventy-five."

Whether my mouth and eyes literally flew open at this point, I don't know. All I can say is that I was stunned. "And immediately there fell from his eyes as it had been scales." Somehow in that instant, the person sitting before me was transformed from a meddling busybody to a lonely and courageous woman doing her best to find challenge and meaning in a life she was not sure she wanted to continue living.

She had told me a few evenings before about the terrible night of her husband's sudden death years ago in a Guatemalan hotel and her subsequent months of anguish and despair, a young serviceman pulling her back from the brink, both literally and figuratively, on a cliff ledge at Machu Pichu—but somehow, I wasn't really listening. "No surprise there," Bonnie would say. Wrapped up in my dealings with Estrada, attending to the injuries and illnesses of my staff, organizing the collecting and processing of the data from each day's netting and censusing, considering how and where the next day's sampling efforts should be focused, and always thinking about Whitmore and Karr and how to find the keys to the deep questions of migrant ecology and evolution their cogent criticisms exposed—I was preoccupied during these evening discussions, to say the least.

The awful events of the last two days, culminating in the revelation of Jonnie's actual age, finally awakened me. You likely will think, "What a jerk. How could he not have known? It was no secret." Absolutely true. All I can say is that until that moment, I did not.

We weren't long with the magistrate. He took Jonnie's statement and said that the defendant would get a couple of years in prison for his misdeeds, which included a previous conviction for horse stealing. He gave Jonnie back her watch and binos, which had been held as evidence, and that was that. We did some shopping for groceries and supplies and headed back to Playa.

After all that had happened, I hadn't the heart to keep up my battles with Estrada and try to protect Lisa and Mike from his henchmen, so we took down our nets and closed up our operation at the station after providing him with copies of our catalogs as per his request.

I set up six nets for Jonnie right outside her Playa room, and on December 5, she made the most important single capture of all my years of work in the Tuxtlas, before then or since—a female Kentucky Warbler, originally captured by us seven years previously on December 28, 1973, as an adult in a net located about one hundred feet from where she was recaptured, still carrying her color bands (Red Left). It is hard to conceive of the thousands of miles flown by this little bird on the biannual trek between her winter territory at Playa and her unknown breeding territory somewhere in the eastern United States.

Jonnie started her long trip home on December 8, catching a ride with Rosie Estrada into Catemaco, then continuing on by bus to Veracruz and plane to Mexico City, Boston, and ultimately, her home in Cape Cod.

My students and I packed up, paid our bill, and said our goodbyes early on December 10, then headed down the road to Catemaco as another *norte* came to drench us and convert the usual half-hour trip to an agonizing two-hour ordeal. We were fine once we got on the paved road, toodling along on a side trip to Xalapa to see Mario and his wife,

Figure 14-4. Kentucky Warbler Red Left, captured at Playa on December 28, 1973, and recaptured by Jonnie Fisk seven years later a hundred feet or so from her original capture point.

Figure 14-5. Mrs. Erma "Jonnie" Fisk.

Isabel; drop off mist nets, specimens, and data for them; and put Isabel Carmona and Sergio on the bus to Mexico City. Then we got back on the main track for the long drive home through Veracruz, Tampico, Brownsville, a brief stop at Welder, and then straight through to Athens, Georgia, arriving early in the morning of the thirteenth.

Jonnie and I maintained a warm correspondence in subsequent years, and she talked on the phone a few times with Bonnie. She mentions her horrendous Tuxtla experience in her book *Parrots' Wood*. To my sorrow, I never saw her again. She passed away in 1990 at the age of eighty-five. For me, Tennyson's *Ulysses*, as quoted at the beginning of the chapter, is an apt comparison. She lived her life with extraordinary courage, continuing to move forward and put herself out there in search of new experiences despite devastating loss.

SCRABBLING ON THE SHOULDERS OF GIANTS

¡Caballeros! ¡Eso no es un mercado!

(Gentlemen! This is not a marketplace!)

—Mario Ramos at the Primer Encuentro
Iberoamericano de Ornitología sobre
Ecología y Comportamiento de las Aves

Tom Lovejoy, arguably the greatest conservation biologist of his era died on Christmas Day 2021. His passing has brought back to me an epochal moment for ornithology and conservation in the hemisphere involving Tom and another giant in the field at which I happened to be present.

Mario Alberto Ramos Olmos, Tom's great protégé, had died suddenly and unexpectedly several years earlier at his home in Washington, DC, in the early morning of September 11, 2006, of heart failure. He was fifty-seven years old. His passing was a staggering loss to the ornithological and conservation worlds to which he had devoted his immense skill, energy, and intelligence for over three decades and a heavy blow to colleagues, friends, and loved ones. Despite his phenomenal accomplishments, there is a sense that he had much left to give.

Mario leaped onto the world stage of ornithology and conservation during the tumultuous concluding plenary session of the Primer Encuentro Iberoamericano de Ornitología sobre Ecología y Comportamiento de las Aves on December 1, 1979, at the professionally tender age of twenty-nine. I recalled that moment in my eulogy for Mario at his memorial at the Washington Universalist Church twenty-seven years later. Tom Lovejoy, caught in DC traffic, arrived a little late to the ceremonies that day and, not realizing I had described that remarkable performance in my comments, proceeded to relate to the gathering the same episode in his own remembrances. Obviously, it was a memorable incident.

Below I present my recollection of the scene. I readily admit that these memories, unassisted by notes and suffused in the sepia tones lent by the intervening forty-two years, may contain some egregious inaccuracies. Nevertheless, as seconded by Dr. Lovejoy and the fact of the current existence of the NOS, the principal proceedings occurred as recounted.

It was Tom, I think, who set the whole train of events in motion, which culminated in Mario's early grasp of the leadership role in Neotropical ornithology and conservation at the Iberoamericano meetings.

As I write these words, it has been a little over a month since Tom's death, hardly enough time to even begin to understand his outstanding legacy. Thinking back to that meeting in Buenos Aires and related events is of some use in these efforts.

My introduction to Tom Lovejoy came at a national meeting of the AOU in the mid-1970s where he delivered a talk summarizing his research on avian species diversity in the Amazon as a postdoctoral fellow of Robert MacArthur at Princeton. A half century later, few recall, I suppose, that to be a MacArthur postdoc at that time was the highest accolade a young scientist could achieve in the field of ecology. Tom could have parlayed that achievement into an outstanding academic career at any of the best universities in the country. Instead, he chose to use his brilliant grasp of the science that provides the foundation for conservation to develop and promote an entirely new field of his own invention—biological conservation.

Tom's unique gift was to be able to turn his sunny smile, dapper mien, and deep understanding of the science of ecology into effective environmental policy. His work in the Amazon gave him unique insight into the two most important factors in world conservation: (1) that habitat loss is the single greatest threat to biodiversity and (2) that the problem is global—all of the various systems in the planet are linked. He began using this understanding immediately on leaving Princeton to take up the office of the vice president of science at the WWF. His job at this organization gave him access to money for research programs he felt he could use to promote conservation. From this position, unlike the administrators in most environmental organizations, he decided to fund work on the science underlying the conservation, which was how I came into contact with him.

Gene Morton had been a close friend and colleague of Tom's in graduate school at Yale, and he approached him regarding support for work on migratory birds in the Neotropics. Tom, I think, saw the value of migrants as a symbol of the way in which ecology throughout the hemisphere is linked and provided excellent funding for a decade for our work on migrants up until his departure from WWF to assume the position of assistant secretary at the Smithsonian. He also saw the conservation value of the theoretical work on island biogeography in terms of Amazonian rainforest conservation—inventing the concept of debt-swap to encourage the preservation of huge blocks of land in Neotropical countries in exchange for debt relief. In addition, he recognized that for science to serve as a basis for conservation, it required a tradition, respect, and understanding of the science, and this understanding had to come from within a country—not be imposed from outside it.

It is a noteworthy aspect of field biology in the United States that our practitioners feel quite at home studying whatever they want wherever they want. We are so used to being able to find money and time to study the bats of Timor or the liverworts of Patagonia that we fail to realize how unique this privilege is. People from the rest of the world find our presence in their lands for these esoteric purposes completely baffling. They always assume that we are there for commercial or political purposes. Tom understood that this lack of a tradition of interest in the natural history of their country was a serious obstacle to any sort of in-country conservation. Until they taught their children to have a concern and raised at least some of them with a desire to find out more, why should they even think about the rich natural heritage of their land?

With this idea in mind, I think, Tom jumped on the idea proposed by Professor Juan Daciuk to organize a meeting of those doing research on Neotropical ornithology from throughout the hemisphere. Dr. Daciuk was a faculty member of the Universidad Nacional de La Plata and not only a leader in avian studies in Argentina and officer of the highly influential Asociación Ornitológica del Plata but a participant in major regional conservation initiatives, such as the Iguazu Falls International Park, shared between Argentina and Brazil.

At the time, most of the ornithological research in Latin America that was done by actual Latin Americans (instead of outlanders) was centered on two countries: Argentina and Brazil. This work was being directed largely by an elite group of white males of European descent (or actual immigrants from Europe such as Swedish-born Claës Olrog in Argentina and German-born Helmut Sick in Brazil). For these men, the hemispheric ornithological scene was something like the old Saul Steinberg New Yorker cover "View of the World from 9th Avenue," with Argentinians and Brazilians metaphorically occupying the Ninth Avenue (New York) seat. It was natural for them to assume that leadership of any hemisphere-wide collaborative initiatives would be headed by them and that a shared old-world cultural aspect would be an important part of any such effort. Which is why the title of this first organizational meeting included the term *Iberoamerican*, stressing those cultural links.

Nevertheless, the Argentinian organizers recognized a hard reality—namely, that the United States was the true gorilla in the room, dominating the hemisphere's politics, economics, conservation, and ornithological work. Therefore, they reached out to the preeminent US players in Neotropical ornithology to participate in helping them organize and fund the initiative. This group, naturally, included Tom Lovejoy.

As I have explained elsewhere above, Tom began providing support from WWF to Mario Ramos and me for investigations of migratory bird ecology in Mexico in the late 1970s. Accordingly, after joining the advisory council for the meeting, he invited Mario and me (and several other US scientists, I believe) to attend the Buenos Aires gathering, November 25 to December 1, 1979, and present papers on our work at WWF's expense, thereby placing the key piece on the board (Mario, that is) for the dramatic events to follow.

The meeting attracted several hundred participants, most of whom, of course, were professors and their students from Argentina and, to a lesser extent, Brazil. There were a number of prominent Neotropical researchers present from the United States and Canada and several from Europe as well (Spain, England, France, Belgium). Most surprising (to me, at least) was the number of Mexican scientists in attendance, including, in addition to Mario, some heavy hitters (professionally) from the leading Mexican biological research institution, the University of Mexico's Instituto de Biología. Such a high level of international participation in what was essentially a regional meeting should have signaled to me that a lot of influential people in the field recognized an opportunity for establishing a new professional organization to promote hemispheric science and cooperation. But it didn't. I just thought that it was an adventure.

Which it was.

The trip down was a nightmare: my credit card information was stolen at the Miami airport (someone looking over my shoulder, I guess, when I called Bon) and there were interminable holdovers en route, resulting in forty sleepless hours before arriving in Buenos Aires—all this topped off by a screaming fight when I refused to pay the cabby and he tried to drive off with my luggage (he had run up the meter by driving around the hotel vicinity a few times). The concierge finally came out to the curb threatening to call the police before the cabby released my bags and peeled out in a huff. I had to reassure him that I was not a habitual deadbeat before he would let me get my room.

Mario and I always roomed together at conferences, and so it was at this one. He had already checked in, coming from Mexico, when I arrived. What I don't remember is how our room became the headquarters where the future of hemispheric ornithological collaboration was planned. Certainly, it had nothing to do with me. I had no illusions as to my stature on the international stage. Nevertheless, it was not long after the conference began that Dr. Daciuk and one or two other senior organizers were getting together with Mario (and me, more or less as an onlooker) in our room to try to figure out how to make this first *encuentro* into an actual organization. I am guessing here, but I think that Gene Eisenmann of the American Museum of Natural History (AMNH) and Tom Lovejoy were instrumental in how this happened. Both men knew Mario well—Tom from his intimate involvement with our research and Eisenmann from his work at AMNH helping Mario with his thesis work on wintering distribution of various migrant subspecies in Central America.

I have already given you some of Mario's history, but it would be well, I think, to give you a bit more at this point in my story.

Mario was born on February 24, 1949, in the Colonia Puebla of Mexico City. During his teen years, his family ran a pharmacy, working hard to become and remain members of Mexico's fragile middle class. As the third of nine children and the oldest male to reach adulthood, he was expected to devote considerable effort to caring for his younger brothers and sisters and to help with the pharmacy. Nevertheless, his parents also expected him to demonstrate sufficient diligence in his studies to qualify for a profession. He showed early promise in the sciences and entered UNAM in 1968, graduating with a degree in biology in 1974. The great ornithologist and student of Mexican birds, Allan Phillips, then a faculty member in the Instituto de Biología at UNAM, was a major advisor for Mario's baccalaureate thesis (1974, UNAM), a study of the birds of the Pedregal, vast lava flows south of Mexico City.

As described in chapter 2, I received a grant from the Welder Wildlife Foundation of Sinton, Texas, to support work on the nonbreeding ecology of migratory birds, with a stipulation by Dr. Clarence Cottam, director at Welder, that the funds be used to support at least one Mexican student. Dwain contacted Allan and asked him for the best student in ornithology in Mexico. Without hesitation, Allan recommended Mario Ramos.

My wife, Bonnie, and I first met Mario at the University of Minnesota's Itasca summer field biology session in July 1973, when Dwain brought him up for a visit. By August of that year, Mario was already at our Mexican field site, working at UNAM's *estación de biología* in the Tuxtla Mountains of southern Veracruz, and living at the nearby *cabañas* of Playa Escondida with Dick Oehlenschlager, where they spent most of the next twenty-two months (until May 1975) gathering data pertinent to the theses of Mario, Dick, and myself.

Assisted by Bon and some outstanding young field biologists, including Chris Barkan, Bruce Fall, and Bob Zink, we accumulated ninety-six thousand net hours in tropical rainforest and second growth during that two-year period, capturing about thirty thousand individuals of more than 150 species. This work forms by far the most significant contribution to the long-running investigation of the birds of the Tuxtlas, a study that began with the work of P. L. Sclater in 1857 and has included contributions from such notables as Alexander Wetmore, Dean Amadon, George Lowery, Robert Dickerman, Allan Phillips, Pierce Brodkorb, Robert Andrle, William Schaldach, Ernest Edwards, and Kevin Winker as well as more than sixty other investigators. A total of 405 species of birds have now been recorded for the Tuxtlas based on this work, 350 of which are documented by specimens. In addition, more than ninety peer-reviewed papers on Tuxtla birds have been produced on topics ranging from ecology, natural history, population dynamics, taxonomy, and conservation to the role of migrants in virus movement. Mario participated as senior investigator, collaborator, or supervisor in a significant amount of this work during his sixteen years (1973–88) as a field biologist before shifting his career into international conservation.

Despite the incredible workload during his graduate research in the Tuxtlas, Mario found time to meet, court, and marry (1975) the beautiful and talented Maria Isabel Castillo, herself a biologist, *profesora*, and UNAM graduate. Issue from this union included three children of outstanding character (Aurora in 1978, Mariano in 1980, and Ameyale in 1981), a coauthored translation of Peterson and Chalif's *Birds of Mexico* (1989), and a number of other shared personal and intellectual achievements.

Mario's brilliant PhD thesis (1983, University of Minnesota) focused on a question raised by Finn Salomonsen and David Lack—namely, "To what degree do birds from a particular part of a species' Holarctic breeding range winter in a specific portion of the Tropical wintering range?" At that time, before feather isotopes, the only way to address the question was band returns or, for those species in which they occurred, regional variations in plumage coloration. Using specimens of eleven species showing identifiable subspecific variation collected in the Tuxtlas, along with hundreds of other specimens from major museums, Mario was able to obtain exceptional insight into the timing, routes, and winter settlement patterns of populations from known breeding regions. This work was quickly recognized by the scientific community as epochal. In 1979, before even completing his thesis, he was offered a plum position at a new Mexican government research organization, INIREB. This organization, based in Xalapa, was created by the renowned Mexican plant ecologist Arturo Gomez-Pompa. Modeled after research institutions like the Max Planck Institute in Germany and the Smithsonian in the United States, research scientists at INIREB had impressive freedom to pursue basic research questions with academic and government support, a freedom that Mario exercised to considerable effect, initiating projects in the Tuxtlas as well as several other regions of the country.

So when Mario showed up at Dr. Daciuk's *encuentro* in Argentina, he was a well-known quantity at the highest levels of the ornithological world, despite his youth, and not just for his excellent abilities as a scientist. In addition to those qualities, Mario possessed a magic that few people possess. It didn't take long in any conversation of substance to recognize that he was special. I think that it was this quality that inspired Tom to bring Mario into the mix early in the discussion of how to use the meeting as the means to build a new platform for collaboration for Neotropical ornithology and conservation.

Still, bringing him into the mix was one thing. Charging to the top of a then nonexistent international professional organization was another. That activity required astonishing vision and forcefulness of personality, and I think these characteristics, in part, helped build toward the dramatic events of the last day of the conference. Nevertheless, I think that the "maestro system" played a role as well.

Dr. Daciuk, as chairman and chief organizer of the current meeting, held the pole position in terms of the formation of any new organization going forward. He was a patrician professor of the old school, and he expected the deference from younger colleagues that his age and professional standing would normally demand. As such, he was used to gatherings in which his status guaranteed a quiet and subservient audience whenever he wished to speak and complete acquiescence with his expressed opinions—but that's not how it went.

The organizational planning meetings were held in the hotel room of Mario and me. They took place normally in the afternoons between sessions and included one or two other senior Argentines in addition to Mario and Dr. Daciuk. I was there by sufferance simply because I lived there. The discussions during these sessions did not follow "maestro" rules, a circumstance that clearly took Dr. Daciuk aback from the get-go. Certainly, Mario was polite, but deference was out of the question. Within minutes of the start of the first meeting, Mario had taken charge. The topic was too important and the time too precious to allow for outdated ideas and platitudes, so shortly after Dr. Daciuk had made a few introductory flourishes, Mario simply assumed the role of chairman, setting out a clear agenda and managing discussion in a productive manner.

The upshot of these gatherings was that by the time the last session of the conference rolled around on the afternoon of December 1, a tentative plan had been formulated for the establishment of a new international Neotropical ornithological organization, and an executive committee had been constituted to assist in carrying out the plan. This final session included all of those who had participated in the meetings. This number amounted to several hundred people, including professors, researchers, and students from throughout the New World and several European countries as well as members of the media and general public. Mario was selected to chair this session.

The gathering took place in a large auditorium. At the front was a stage. Members of the recently appointed executive committee were seated at a long table, with Mario occupying a central position. I know that the members of the executive committee included Dr. Daciuk, Gene Eisenmann, and Tom Lovejoy. I'm not sure about the others—probably Drs. Olrog and Sick as among the most eminent attendants at the *encuentro*. The agenda had Dr. Daciuk leading off, thanking folks for their attendance and mentioning his hope that the actions of the executive committee would lead to a continuance of a Neotropical ornithological organization. Then Mario would take over, and time would be given to hear and discuss proposed resolutions.

At the time of this meeting, resolutions were a standard part of the conclusion of most professional conferences. These resolutions, which would be part of the published proceedings resulting from the meeting, would give professional weight to topics of importance for both the scientific world and the general public. An example might be something like "The members of the Primer Encuentro Iberoamericano de Ornitología sobre Ecología y Comportamiento de las Aves resolve that the use of DDT as an agricultural pesticide should be prohibited due to its action as a lethal poison for members of many aquatic environments." The idea was that time would be allowed for ten or fifteen such resolutions to be put forward by the audience, followed by some discussion, and then there would be a break of fifteen minutes or so before reconvening for the final session of the meetings, wherein the executive committee would announce their selection of the person to chair the next meeting of the nascent organization, and that person would explain plans for how and when that meeting should occur.

Unfortunately, this anodyne-sounding agenda didn't unfold quite as planned. The first few resolutions went more or less according to Hoyle. However, it didn't last. A resolution was proposed where half the professors and their students at Argentine universities were in favor, while the other half were opposed. As soon as this resolution was put forward, a vehement negative response was voiced in opposition. Mario tried to calm the discussion by asking folks to limit the amplification and length of their opinions, but that simply seemed to increase the volume, lest some might not have the chance to express themselves. Soon what began as a skirmish deteriorated into all-out war. Pandemonium ruled. Entire sections of the auditorium were filled with people standing, screaming, and gesticulating. The executive committee members were looking at one another in complete stunned bemusement. Into the breach stepped Mario. Removing his shoe in Khrushchev fashion to use as a gavel, he pounded the table and roared the words quoted at the beginning of this chapter: "¡Caballeros! ¡Eso no es un mercado!"

Quiet ensued.

Now that he had control, Mario said that the resolutions session was terminated. People were asked to submit their resolutions in written form to Dr. Daciuk's *encuentro* committee. He then thanked those present for a frank discussion and proposed that we reconvene for the final session in half an hour.

During the break, I went up to Mario, who was discussing things on the stage with the committee. "Have you ever heard of Robert's Rules of Order?" I asked. He answered in the negative. "Well," I said, "it's a set of guidelines for how to conduct a meeting. Pretty much every meeting of any kind in the US from Cub Scouts to Congress uses some form of these rules. You really need them to keep things going along smoothly." I then explained that just establishing some guidelines might help the current situation—like no one talking without the chairman's permission and limited time for each person's comments. He said he thought it sounded like a good idea, and so when the session reconvened, he laid out a few of the most basic tenets.

After all that, the committee's recommendation that Mario be the person charged with organizing and holding the next meeting of the new organization, to be called the Segundo Congreso Neotropical, was pretty anticlimactic. To have selected anyone else would have been comparable to taking the chair out of the lion tamer's hand—with the lion still pacing ominously in front of him.

And history has proven them right. Mario took full responsibility for organizing and funding the next meeting, which was held in Xalapa, Mexico, in 1984. At that meeting, all business sessions were conducted according to Robert's Rules, and a new organization was established, the NOS. Bylaws for the society were established, and officers were elected, with Mario as president (a post he held for the next fourteen years until he was certain that his replacement had sufficient organizational will behind him to hold the society together into the future). A journal for the society was planned and an editor selected.

With Mario's passing, both Tom and I, independently, realized that Mario's actions at the Argentinian meeting had signaled the arrival on the international science and conservation stage of a marvelous new actor. The man and the moment came together then.

The quality of Mario's research and his obvious abilities sparked a meteoric rise for him in international scientific and conservation circles from early on in his career: 1980, elected president of the Mexican section of the International Council for Bird Preservation; 1980, founded the Mexican conservation organization, Pronatura; 1982, invited to join the Committee of the International Ornithological Congress; 1984, founded the NOS; 1985, became an elective member of the AOU; 1986, invited to present his thesis findings as one of five plenary speakers at the Nineteenth International Ornithological Congress, Ottawa, Canada (Ian Newton, Jacques Blondel, F. Cooke, and Peter Berthold were the others); 1986, selected as the Senior Conservation Fellow for WWF; 1992, elected fellow of the AOU. However, his departure from INIREB for WWF in 1988 effectually ended his work as a field biologist, and the last time we worked together in the Tuxtlas was radio-tracking territorial Wood Thrushes in January 1988 at the *estación de biología*. Nevertheless, Mario maintained his professional interest in ornithology until his death, serving as the president of the NOS from 1984 to 1998, attending national and international ornithological congresses (such as Beijing in 2002, the last time we roomed together), and publishing in the ornithological literature. As recently as August 2006 (a month prior to his death), we had discussed the possibility of returning to the Tuxtlas to continue our work in a year or two once he had finished up his day job as senior environmental consultant for the World Bank.

Mario had an uncanny facility to sift vast amounts of information, identify the key issue, and chart a practical course of action—skills that stood him well in his profession. However, the clearest evidence of Mario's impressive gifts was most readily observed in a public forum—any meeting where important issues were at stake. He could argue cogently and forcefully in three languages (Spanish, English, and Portuguese), and I rarely saw him fail to make his point and achieve his goals, regardless of the audience size or composition, from a few *campesinos* in a Tuxtla *ejido* to several hundred rowdy colleagues at a Buenos Aires auditorium. These abilities, along with his deep commitment to conservation, were a key part of what made Mario a man of world stature who had a profound influence through his fifteen years of work at the World Bank and its Global Environment Facility (GEF). However, such labors, though stunning in their demands and crushing in their toll, leave remarkably little in terms of identifiable personal legacy, and I think it is as Mexico's greatest ornithologist and conservationist that he will be best remembered. His publications authored, students trained, organizations fathered, and wildlife preserves established assure him a lasting name in this arena. And yet his membership in the pantheon seems somewhat odd to those of us who knew him as a comrade, sharing soggy *tacos gringos* (tortillas and peanut butter) in the cold, never-ending rain of a Santa Marta cloud forest. True, his strength, energy, focus, and patience at times appeared superhuman in this environment, but we didn't think or worry much about his international standing. We knew him then mainly as we remember him today: as a dear friend, loyal, dependable, and thoroughly enjoyable to be with regardless of the circumstances, and now as one who has left us way too soon.

I have been gifted in my life to work with some of the most remarkable people of our time. It is an incredible thing, but when you meet them, you recognize them immediately. Physical appearance has nothing whatsoever to do with it. Sparks fly, and while I make no claim to an ability to perceive auras, one can sense their special nature immediately. Tom Lovejoy and Mario Ramos were such people.

CHAPTER 16

KINGSVILLE

Kingsville isn't the end of the world
But you can see it from there

—Anonymous (from a T-shirt)

Ronald Reagan was elected the fortieth president of the United States in November 1980. This event, I was quite certain, marked the end of our Georgia idyll. Funding for the positions of my colleagues and me at the university came from a large grant ($200,000) under the auspices of section 6 of the Endangered Species Act of 1973, established by Congress to provide money to the States to find out what species within their confines needed help. Reagan had made no secret of his disdain for such wasteful spending during his campaign, and I was certain that our money would disappear like spit on a griddle were he elected. Indeed, I had told my Republican colleagues, "Go ahead. Vote your values over your pocketbook." I don't know whether they did or not, but enough people did to put an end to such frivolous use of public funds. Reagan's interior secretary, James Watt, quickly killed as much endangered species activity as Congress and the courts would allow, which at least put paid to our little research group in Georgia.

By June 1981 it was clear that my family and I would be back out on the streets unless I could find another job somewhere. So I sent out an SOS to my various professional friends around the country, asking them to let me know of any positions that might be coming open. This cry for help yielded immediate results from one of my Welder mentors, Caleb Glazener, who had assumed the directorship on Dr. Cottam's death. Mr. Glazener wrote to say he had heard that Texas A&I University in Kingsville, Texas (about sixty miles south of Welder), had gotten a million-dollar gift from the King Ranch to establish a natural history museum on their campus. Accordingly, they were looking for someone to be the curator of natural history to head it up. He said he knew the woman who was in charge of the search, Mrs. Jimmie Picquet, and would be happy to contact her on my behalf if I was interested.

I mentioned back in chapter 1 that natural history museums have long served as the scientific foundations for field biology studies here in the United States and elsewhere. The opportunity to lead such a venture with what sounded like significant financial backing was both challenging and exciting. In addition, the Texas Gulf Coast is one of the great migratory bird pathways in the world, offering significant research possibilities that we had barely begun to investigate during our time at Welder. Also, from a personal perspective, Bonnie and I had many friends in South Texas and thought we knew the place pretty well.

With these thoughts in mind, I contacted Mr. Glazener to let him know that I was interested and sent an application for the position to Mrs. Picquet. Things moved fast after that. Within a month, I was invited to Kingsville for an interview, during which I was offered the job, and I accepted. "No way I would consider another candidate after hearing from Caleb," Jimmie said. A couple of weeks later, Bonnie and I flew out to see if we could find a house and bought one at 628 West Lee, about six blocks from the museum building. We were settled in Kingsville by early August.

Jimmie Ruth Picquet was fifty-four years of age when she wrangled the King Ranch gift from her perch as director of the Conner Museum. She was the prototype of the Texas "new woman." Valedictorian of her high school class in Bishop, a town a few miles north of Kingsville, she had gotten a degree in music from the University of Texas before settling down as the wife of a local farmer to raise three sons. That duty successfully accomplished, she went back to school at Texas A&I and obtained her master's degree under the tutelage of legendary history professor John E. Conner, who had served in that capacity since the university's founding in 1925 (originally as South Texas State Teachers' College). Professor Conner had begun collecting materials for a historical museum shortly after he began his tenure at the school and by 1929 already had a campus room set aside as the John E. Conner Museum of History to house these memorabilia. Jimmie became Professor Conner's protégé, and in 1972, she assumed directorship of his museum.

It has been my experience that many Texas professional women share similar behavioral traits: brash, overbearing, sharp-witted and sharper-tongued, and lacking the slightest hint of political correctness. I have a theory that this is a cultural phenomenon derived from earlier generations when pioneer women often were left completely on their on to manage the ranch, fend off the rustlers, raise the kids, and fight the Comanches while their men were at the railhead in Kansas City taking a little R&R. Whatever the truth of this assessment might be, Jimmie was the archetype for me. She wore her "big-girl panties" at all times and expected other women to be similarly armored in their dealings at the workplace. She called her secretary "chickie babe" and brooked no whining from anyone, regardless of gender.

In any event, Jimmie was used to being, or at least acting as, the smartest person in the room, and it did not take her long to transform the sleepy institution of the Conner Museum. The South Texas gentry was a small and exclusive club in those days, and Jimmie knew them all, including the Klebergs, who were the leading members of the King Ranch conglomerate. She soon parlayed her moxie and contacts into making her little museum a campus political force to be reckoned with.

The Klebergs had a long-standing love for, and interest in, South Texas wildlife and natural history, and in the early 1980s, they decided to put significant money into Texas A&I to build programs in those areas. The College of Agriculture was the obvious choice to receive the million-dollar donation from the ranch's Caesar Kleberg Foundation to develop a wildlife research institute. Less obvious was the choice of Jimmie's organization as the recipient of another million to build a natural history museum and associated academic and research programs.

Natural history museums, usually in partnership with major universities, historically have provided the nuclei for much of the best work in field biology. Unlike their counterparts in other disciplines, and although crammed with dead things, they are the living heart of investigation into how things live in the real world.

When Jimmie won the King Ranch grant, her idea was to transform the museum of which she was director—basically a couple of rooms crammed with old branding irons, barbed wire, arrowheads, potsherds, and the like—into a thirteen-thousand-square-foot display area filled with the most beautiful and inventive exhibits that the country's best diorama makers could provide. My colleague at the Georgia Museum of Natural History, Lloyd Logan, had quite a bit of experience in this aspect of natural history museum design, and I was able to convince Jimmie to hire Lloyd as well. Within a couple of months after the arrival of Lloyd and myself, we had hired a design firm and were well underway.

A year and a half later, my field colleagues, Hal Ham and Lloyd Logan, and I had done most of the collecting required to fill the exhibits but had accomplished little in the way of research. In addition, thanks to various misunderstandings and snafus involving our contractors and subcontractors, the new natural history wing threatened to be as dead as the stuffed javelina that greeted visitors to our little shop. It was at about this time that I went to a faculty party at the house of the president of the university, Bill Franklin. There, I ended up talking with Dick Meyer, academic vice president and number-two man at the university. He asked how things were going, knowing already about our difficulties with our design firm. I explained that these issues were not the fundamental problem in my opinion. I explained that a natural history museum in most places was primarily programs, not displays, and that if the concept was to have a future, it needed to be tied to an academic arm of the university rather than a stand-alone operation as envisaged by Jimmie. I suggested that separating the natural history museum from the history museum and aligning it with faculty from the biology, geology, and wildlife departments through joint appointments of their faculty as curators would be a better way to go. He said that I should write up a proposal suggesting how we might proceed, which I did. However, when he showed this to the university president, Dr. Franklin wanted to know if Jimmie was aware of this document. Meyer admitted that she wasn't. So Franklin called Jimmie, informing her of my betrayal and suggesting the termination of my contract at the end of my second year, which is what happened.

Things looked pretty bleak in the summer of 1983. I was soon to be jobless, and Bonnie had to go back to work immediately (as a visiting home health nurse with a five-county responsibility—an area the size of the entire state of Massachusetts). Jimmie, I think, felt a bit trapped by the president's actions. In any event, despite my firing, she recommended me for a position with the other King Ranch–sponsored organization on campus, the Caesar Kleberg Wildlife Research Institute (CKWRI). Sam Beasom, the director, hired me in a low-level postdoctoral position for $18,000 a year to study White-winged Doves, a game bird native to the Rio Grande Valley. This salary represented a considerable reduction, but at least I had a job.

Thirty-eight years on, it is interesting to compare the John E. Conner Museum, where I was first employed at A&I, with the CKWRI, my second employer at the university, as they stand today—kind of an academic version of the morality tale "The Three Little Pigs." Both organizations were gifted in the late 1970s with million-dollar grants from the King Ranch. Jimmie focused her funds on the development of a display facility; CKWRI used theirs to hire a quality academic research staff and begin building an endowment. Today the Conner Museum is four rooms instead of two—adding two rooms of natural history displays to its original two rooms of historical displays—while CKWRI has become one of the premiere wildlife research institutions in the country, if not the world, with a hundred-million-dollar endowment and outstanding faculty.

THE LA PENINSULA GAMBIT

Regardless of our differences concerning how a natural history museum should be organized and operated, my boss at the John E. Conner Museum at Texas A&I, Jimmie Picquet, never interfered with how I thought I should spend my time as natural history curator. She gave me complete freedom in that regard, as did my subsequent employer at the CKWRI, Sam Beasom, after my move across campus in the summer of 1983. As a result, I was able to continue my work on migrants at our study sites in southern Veracruz throughout this period.

Chapter 14 dealt with our return to my Playa Escondida study site after a five-year hiatus, during which time the forest on my main study site had been converted to a mixture of cornfield, pasture, thickets, and remnant forest. This conversion allowed us to compare Nearctic migrant species composition, population size, behavior, and site tenacity before and after the conversion. We found that in forest, members of these species held territories or joined mixed-species flocks and remained in a single small block of forest throughout the winter season, from their time of arrival in September or October to their departure in March or April, and returned in subsequent years. However, members of these same species captured in scrubby pasture, cornfield, or hedgerow showed no site fidelity. They disappeared within one or two days after capture. The total number of these individuals captured might be higher in the second-growth habitats over the course of the season than in forest, but there is a dramatic difference in the birds' behavior. They are not territorial and show no within-or between-season fidelity. These are the wanderers predicted by the northern home theory. The reason that these birds are wanderers seems clear enough: they cannot find a stable food source to allow them to remain in one place, unlike those in neighboring forest. Therefore, it is apparent that there are at least two different strategies potentially followed by migrants wintering in the tropics: one in which they fly southward to occupy a preferred habitat where they remain through the winter before returning to their breeding grounds, and one in which they fly south and wander throughout the winter period in search of temporarily superabundant food concentrations. Naturally, one would expect that this wandering group would be composed largely of young birds heading south for their first winter season. After all, they have never been south before, and as we showed during our Welder studies of transients, they are on their own during their migratory travels with no one to guide them. However, consider that vast amounts of forest and other primary habitats have been altered or destroyed and replaced with very different kinds of habitats, like pasture and cornfields. These changes probably force many adults into a wandering mode as well, since the preferred habitats where they wintered in previous years no longer exist.

The fact that we captured individuals of forest-related migrants in at least some kinds of marginal second-growth habitats in February and March, late in the winter season, demonstrates that the wandering strategy can work. You can use it and have a chance of surviving through the winter to migrate north in the spring, though your probability of success may be quite different from your sisters and brothers occupying more stable and productive environments. This observation contains my critic R. C. Whitmore's most cogent argument, which he might have phrased as follows: "OK, so you have some birds holding territory in forest. So what? Lots of birds of the same species don't, and they seem to do just fine. Unless you can show fitness differences between members of the two different behavioral groups, you really don't have anything."

So it all comes down to fitness. *Fitness* is population genetics jargon for the number of offspring contributed by an individual to the next breeding generation. It is how evolution works: the more individuals you contribute to the next generation, the more of your genes you pass on.

If you thought about it, you would have to concede that we had pretty good data on the likely fitness of our winter territory holders, at least for a few species like the Hooded Warbler. We knew that they suffered low mortality during the winter season—zero, in reality, for hoodeds on our study sites whose territories weren't destroyed by cutting. Also we had high return rates for the next season. True, we did not know how many young they had raised up north, but we at least knew they had survived to come back down the next winter season. Jonnie's Kentucky Warbler was our Exhibit A in that regard, a seven-year-old snowbird. Nevertheless, getting winter survivorship data on wanderers presented a different kind of problem altogether. They didn't stick around for more than a couple of days and were never seen again. It was logical to us to assume that these birds would have a higher mortality rate, and hence lower fitness, than our sedentary territory holders, but that was just supposition. How could we test it?

In 1982, Mario, Gene Morton, and I put forward a major proposal to the WWF that we hoped would allow us to do just that. The proposal we submitted was similar to the one Gene and I had submitted to the NSF back in 1976. In its essence, the purpose was to continue the investigation of Nearctic migrant and resident species in undisturbed tropical forest as compared with various forms of open and second-growth habitats. We would use the same methods of banding, recapture, and observation as we had already been using. But we would add one other technique: radio-tracking.

Radio-tracking involves the attachment of a small transmitter to the back of a bird and subsequently locating it (the transmitter and bird) using an antenna and receiver. I have mentioned that Jim Karr's critique of our earlier proposal to NSF had killed it. Most of his negative comments were directed at our proposed use of radio-tracking, and in truth, they were not without merit. At the time, it was a relatively new technique that had been used mostly on medium-sized or large mammals or birds in relatively open, temperate habitats. No one had used it to follow small birds in tropical rainforest, where the weak signal from the necessarily small transmitter likely would not travel far.

Figure 17-1. Kevin Winker attaching a radio transmitter to a Wood Thrush.

Jim Karr, fortunately, was not a reviewer of our WWF proposal. Nevertheless, the scientists who did had similar concerns regarding our NSF proposal. Fortunately, we were able to convince them that we would get it to work. We also made the case that it was the only way to get any data on the fitness of wanderers, which is as true now as it was then.

From a conservation perspective, in order to make the case that the destruction of tropical forest had a negative effect on migratory bird populations, we had to test whether the enforced shift from a territorial mode of existence in forest to a wandering mode in marginal habitat lowered individual fitness, resulting in declines in the number of migrants returning to breed in the United States. From the point of view of testing our theory, it was important to find out whether tropical forest was the true home for migrants to the temperate zone.

There was a third aspect, however, to the proposed work. In addition to the testing of ecological, evolutionary, and conservation hypotheses, which had been sufficient to warrant funding in previous proposals, WWF now required a human development aspect. In other words, we had to tie our scientific investigations to some form of assistance to the *campesinos* on whose land we were working.

The best approach, from my perspective, would have been to lease the land used for our studies, just as a farmer or rancher might lease an additional crop or rangeland from a neighbor. Too simple. Organizations like WWF (after Tom Lovejoy's departure) wanted projects that not only gave some support to the local people but helped provide them with long-term methods of using the land that were not destructive of its conservation value—a new, complicated twist to the old Chinese proverb "Give a man a fish, feed him for a day; teach him to fish, and you feed him for a lifetime," to which you now must add "Teach him to fish responsibly, and future generations will be fed as well, and there will still be an ecosystem."

Mario, Gene, and I had no background or training in this kind of work, but Mario's parent organization, INIREB, did. In fact, it was part of their mandate as established by their founder, Arturo Gomez-Pompa. In addition to the scientists like Mario at INIREB, there were economists, agronomists, fisheries people, and even sociologists trained to work as teams to develop new ways for the millions of members of Mexico's subsistence economy to feed themselves and their families without destroying the natural biota. So, accepting that we were complicit in ignoring our scruples to accept the funds, we did what we had to do to get the grant, incorporating the INIREB teams into our proposal, even though we knew that despite the fact that we were the PIs (principal investigators), we would have no real control over what these folks would do—or, as it turned out, would fail to do.

The grant was for $100,000 over a six-year period, not a lot of money for a field/development project in rural Mexico, but all three of us had jobs, so no salary had to come out of it. It seems appropriate to mention here that since the expiration of my Welder grant, nearly all of the funding for Mario, Gene, and me in the Tuxtlas had come from the WWF. This support was due almost entirely to Tom Lovejoy, then vice president of science at WWF. After Tom left to become assistant secretary at the Smithsonian in 1987, there was no more money for pure research available from WWF or hardly anyplace else other than NSF. As I have mentioned, even this WWF grant was tied to a quasi-development project. Tom had science chops of the first order: a PhD from Yale and a postdoc with MacArthur studying tropical avian diversity in the Amazon basin. He was a major architect of the US debt-swap for conservation initiative throughout the hemisphere; a developer of the concept of biodiversity; a critical supporter of studies of habitat island size in the Amazon, pushing for the setting aside of large forest blocks by the Brazilian government; and a cofounder of the field of conservation biology. Arguably, he was the single most influential scientist on US government international science and conservation policy for the past forty years. He should have been secretary of the Smithsonian rather than those chosen by the Board of Regents over that period (my personal opinion based on service under the aegis of four of them). No scientist of his stature has filled that position since S. Dillon Ripley retired in 1984.

The community development factor complicated our search for a suitable study area in the Tuxtlas quite a bit. Where were we to find a community that owned rainforest, was willing to keep it in rainforest for the duration of the study, and would participate in ad hoc community development projects? Mario and his students did all of the prospecting, and in August 1982, he let me know they had found a place east of Coyame he thought might be suitable.

Home base at the time of this work was Texas A&I University. Accordingly, my graduate student Kevin Ballard (who had followed me from Georgia to Texas) and I set out early on the morning of September 10, 1982, from Kingsville for the Tuxtlas, driving two vehicles, the "white whale" (an ancient, three-quarter-ton pickup belonging to Texas A&I) and a big, beautiful 1979 Chevy Blazer purchased with WWF grant funds, which we were going to leave at the new study site. The plan was to meet Mario and his crew in Catemaco at ten in the morning of the eleventh, then go on to the new site. We ran into problems immediately at the border. Due to plummeting oil prices, the Mexican economy was in free fall; the banks were nationalized on the sixth, and the value of the peso went from twenty-three to seventy-five pesos to the dollar in a matter of hours. Unfortunately, angst infected the *aduana* as well. When we went to obtain our visas, instead of taking a nice tip and waving us on through, they wanted to know what we were doing with all the supplies and scientific equipment filling our trucks. I told them I was conducting a joint research project with Mexican scientists in southern Veracruz. They shook their heads and took us to the *jefe* (chief), who shook his head and sent us to the Mexican consulate in Brownsville, where the woman in charge, Señora Rojas, shook her head too and refused to give us visas. She said that as scientists, we had to apply through the Corpus Christi consulate via the State Department—about a six-month wait. I explained that we were to meet our Mexican counterparts in Catemaco at ten the next morning. She said, "Well then, go as tourists." This was good advice that I have followed on more than one occasion since. If you don't get the answer you want from one person, try another and/or another tale. In fact, I once had to try four different border crossings before I could figure out the right combination of explanation and customs officers that would let us cross.

Unfortunately, however, this tactic did not work this time. When we came back to the *aduana* and claimed that we were no longer scientists but tourists, they recognized us, shook their heads, and sent us again to the *jefe*, who sent us to the sub-*jefe*, who explained that we absolutely had to have the science visa. It was a new rule. And it had to come from the consulate. I asked him to call Señora Rojas and explain this to her, which he did. Then we went back to Señora Rojas and she finally issued the visas, although not very happily. We then headed to the border for the third time, got our visas stamped, and moved on to the usual customs inspection for the importation of smuggled goods.

The NRA is not lying when they say that gun ownership is difficult in many countries. Certainly, that was true in Mexico in the early 1980s, when it was problematic for a private citizen in Mexico to obtain or own a gun. I did not need a gun for my work, but Mario did, and he asked me to bring a shotgun and shells down with us. Normally, shotgun import is not difficult. Hundreds of Texas hunters do it every year, mostly for shooting White-winged Doves. You need (or needed at that time) a letter signed by your local sheriff or chief of police on official cop stationery stating that you were a law-abiding citizen with no criminal record. Presenting this at the consulate, they stamped your visa, noting that you have imported a shotgun. The problem is that you have to have the gun with you when you leave, and Mario wanted it for the duration of the study, six years. The obvious solution was to smuggle it in. This procedure was no problem normally, given the traditional casual wave-through by *aduana*. So I packed a shotgun and five boxes of shells under a ton of stuff in my new Blazer and trusted to luck.

As it turned out, we needed quite a bit of it.

The initial inspection went as expected, with little examination of our loads, and we thought we were free and clear. What we did not realize was that the rules had changed. Now instead of just one customs inspection, there were three. Also, times were tense. The army was on full alert, and performing its own ad hoc inspections along major highways, looking mostly for guns. Blithely unaware of these challenges, we went on our way until we hit the second customs stop ten miles from the border. They had us open up the homemade wooden camper top of the white whale, gave it a look, and sent us along. We went on for another hour or so and then were stopped by the army. They asked us if we had any guns. We gave them some cakes and Cokes and said, "What do you take us for, idiots?" and they waved us through. About an hour south of San Fernando, we came to our third customs check, and this guy was a problem. He said that we had to pay taxes on the equipment and supplies. I said this was absurd, that the stuff was for a scientific study in collaboration with Mexican colleagues, and if we were supposed to pay taxes on it, why hadn't the other two customs inspections charged us? He got mad, made us unload everything, nearly finding the shotgun and shells, but finally let us go.

The fifth stop was the scariest. We were hours behind schedule and, with no way to let Mario know, were pushing on through the night to try to keep our 10:00 a.m. appointment. At about two in the morning, we were stopped again by the army. This time the squad leader, a lieutenant, came to the window and, despite the cakes and Cokes donation, decided that both vehicles would have to be entirely unpacked and searched for guns. He was probably bored and maybe felt that his boys needed some work. In any event, he stepped back and motioned for two privates to oversee the unpacking and inspection. There was nothing for it, so we took everything out until the only things remaining in the back of the Blazer were the shotgun and shells.

At this point something occurred that I will never understand. The soldier looked right at the gun and shells and said, "OK. You're good to go." It was like that moment in *Star Wars* when Obi-Wan tells the stormtrooper, "You don't need to see my identification." Only I hadn't said a thing. Sometimes the Force is just with you, regardless.

Worse was to come. Much worse.

A couple of hours later, we were flying down the coastal road just north of Nautla at about sixty miles an hour, me in front in the whale, Kevin following in the Blazer, as it was just starting to get light. I checked my rearview mirror, saw Kevin tracking right behind me, checked the road ahead, and then the mirror again to find the Blazer had vanished! It took me a second to realize Kevin must have fallen asleep and driven off the road. The probable reason I couldn't see the Blazer was that there was a ten-foot embankment over which the truck likely had gone. I pulled the whale over to the side of the road and sat there.

That was a bad moment. I was pretty sure what I would find and wanted to be ready to handle it, just like they teach you in basic training in the army. If he's alive, stop the bleeding, clear the airway, move him into the truck, treat for shock, and get him to the nearest medical facility as soon as possible.

Taking a deep breath to gather myself, I turned the truck around and headed back. Sure enough, there was the Blazer. It had gone over the edge, down into a wet meadow, and traveled a few feet, side-swiping a concretion about two feet in diameter with its right side and both right wheels before coming to rest in a foot or so of mud.

Scrambling down the bank and across the marsh to the driver's side door, I was amazed and delighted to find Kevin unhurt, if a bit dazed. The Blazer was not too bad either. The right side was a bit caved in, both tires on the right side were blown, and the contents looked like they had been sorted by an incautious gorilla, but other than that, it seemed operable. I have driven that road a lot since and looked to see how many places he could have driven off the road at speed and survived. The answer is not many. He didn't even have on his seat belt.

All things considered, we had much to be thankful for. Nevertheless, even Pollyanna would have had to admit that our prospects looked dim. But this was Mexico, and within about two minutes, a bus filled with workers on their way to a job saw us gazing forlornly at our impossible situation and stopped to see if they could help. I said I didn't see how, but they all piled out anyway—maybe fifteen men or so. There was much discussion, but they figured that if we ran a cable from the whale to the Blazer, put the Blazer in four-wheel drive reverse, and all pushed, we could get it out. I was skeptical, but what the heck? We gave it a try, and out it came! They wouldn't take any money, *hermanos del camino* and all that, and headed off to work, leaving us with our wounded but extricated vehicle back up on the shoulder of the road.

We set to work to get the wheels off so I could take them to a *vulcanizadora* to get the tires replaced. The front one came off, no problem, but we had to drive a little bit with the nuts off to get the rear one to come loose. By ten in the morning, we had the wheels back on with their new tires and were on our way to Catemaco. We got there at about four in the afternoon, six hours late, but alive and mostly well. Mario and his crew were waiting, and so we took off right away, our destination Ejido La Peninsula de Moreno, located about ten miles off the main road in the heart of the Santa Martas.

I need to pause the narrative here for a moment to explain why we chose La Peninsula, one of the more logistically difficult places imaginable, as a home base for our fieldwork for the next six years. As I have mentioned, lowland tropical forest was rare in eastern Mexico by the time we began our studies in the early 1970s. The Tuxtla Mountain region of southern Veracruz held the largest remaining blocks, mainly because of the rugged terrain and high orographic rainfall. Indian agriculture-based civilizations, beginning with the Olmecs, a thousand years before the birth of Christ, flourished in the lowlands of southern Veracruz surrounding the Tuxtlas up until the Spanish invasion in the early 1500s. However, these cultures did not include the Tuxtla region, home to its own indigenous people of largely forest-dwelling hunter-gatherers. These people are called Popoluca, a Nahuatl (the Aztec language) term meaning "gibberish" and used generically for any indigenous people who didn't speak Nahuatl. Indeed, the term is used to this day for a number of different tribal groups from different parts of the country that may or may not be related. The Tuxtla group is linguistically related to the Zoque, an indigenous people whose origins appear to be in northern Chiapas. Popoluca still live in the Tuxtlas; Angel (of Playa) and his family were Popoluca.

The founding of the Tuxtla towns of San Andrés and Catemaco dates back to the early 1530s, but not much clearing of the forests was done outside the vicinity of these settlements even up until the 1930s. However, with the passage of the agrarian law in 1934, state lands, which included most of the Tuxtlas (no recognition was given to the Popoluca), were made available for settlement under both the *ejido* and *colonia* systems, and colonists began to establish ranches and settlements deeper and deeper into the interior. By the 1950s, about half of the lowlands had been cleared of forest and settled, and by the early 1970s, when we arrived, about two-thirds had been cleared, which included nearly all of the readily accessible lowlands.

Ejido La Peninsula de Moreno was chartered in 1968. Perched precariously on a relatively flat shelf of land from which slopes rose steeply on the wrong side of the Coxcoapan River—*wrong* meaning on the opposite side of the river from where the logging road, which represented the only contact to the outside world in general and possible markets for *ejido* goods in particular, ended. There was no bridge, and none likely. They had to cross the river to even begin the ten-mile walk to town, and for significant parts of the year, it was impassable. Presumably, they had no choice in this because they did not own the land where the road ended. The appeal of this desperate community to us was that over half of their land remained in lowland rainforest, one of the few remaining such sites in the entire region.

The logging road that ran from the main road at Coyame on the banks of Lake Catemaco to La Peninsula was not maintained as a road for normal vehicles. In fact, it brought to mind an observation from May 3, 1827, regarding the road from Tampico to Mexico City by the naturalist Jean Louis Berlandier: "That which the muleteers are accustomed to call a highway (*camino real*) is a badly laid out path from which one can stray without being aware of it" (translated from the French by Sheila Ohlendorf). No one in La Peninsula owned a vehicle. There were small ranches along the route, and some of these guys had trucks, but traffic certainly was limited. As a result, it was often an adventure to make the drive, and so it was on my first trip in on the afternoon of September 11.

Our caravan included me in the Blazer, Kevin driving the whale, and Mario with his student Sergio Barrios (who had helped me with Jonnie) and assistant Jacinto Hernandez Peña, leading the way in an INIREB jeep to which was attached an eight-by-fifteen-foot house trailer, also INIREB property, which was to be left as housing for us and whatever students were doing work there. It was only ten miles from Coyame to La Peninsula, but it was not an easy drive: we had to inch our way past washouts, across a couple of arroyos, and across trackless, open pastures. It was while negotiating the latter that I blew out the same two tires blown out earlier in the day by striking a sharp rock hidden by the grass. We left the Blazer to be dealt with tomorrow and crawled on to our destination on the banks of the Coxcoapan, directly across from Ejido La Peninsula, a village consisting of forty-five huts for *ejidotarios* and their families, a concrete building that served as school and meeting hall, and a wooden structure for communal storage. We set up camp, licked our wounds, and went to "bed" (a pad in the back of the whale for me) for the first time since 4:00 a.m. two days previously.

There were four objectives for this initial visit to La Peninsula: to negotiate a long-term deal with the *ejido* for the use of the forest for our study, to get the trailer set up as housing for ourselves and our students during our research visits, to construct a rope bridge that would allow us to cross the Coxcoapan whenever we wanted to regardless of its flood stage, and to set up a grid of mist-net sites on our forest study area.

Mario had talked with Pascual, the *ejido* president, and Agostín, the *secretario*, earlier to present our proposition, which, at this stage, was simply that we wanted to do a long-term study using their forest; in return, we would work out some sort of development assistance for the *ejido*. It was pretty sketchy, but we didn't know much more at the time.

The *ejidotarios*, all forty-five of them, had met sometime before our arrival and given their tentative approval for us to begin setting up the study area, pending further elucidation of what the "development assistance" might mean. Consequently, we set out on the morning of the twelfth to locate and map out a forest study area and begin building a rope bridge. The weather held through the thirteenth, but rain moved in during that night, and it rained hard every day thereafter until our departure on the twenty-fourth. Well. It *was* the rainy season (May to October). It did give us some chuckles, though, that Agostín had told us that the river was too high to cross on foot only "three or four times a year." Maybe. That was not to be our experience during the five years of our work to come in the area, but you have to love the optimism.

It took us a couple of days to get the rope bridge working. This structure involved two trees, one on either side of the main channel, each with a ladder to a platform twenty feet above the river where a rope was connected to a pulley and harness, with a return rope attached, like a zip line: crawl up the ladder to the platform, hook yourself into the harness, make sure the return rope is clear, and off you go. Simple and reasonably safe—unless you happened to hang up the return rope on something. But nothing's perfect. We completed this chore by the fifteenth, which, as it turned out, was cutting it

Figure 17-2. Mario testing the rope bridge at La Peninsula.

close; the river became impassable on foot by the seventeenth and remained so for the next week.

Our first villager to use the line was a ten-year-old boy, Roberto, whom Mario and I discovered cowering and shivering in the dark under a piece of plastic on the wrong side of the river on the evening of the seventeenth. I took him to the platform, hooked him up with me, and we zipped across. Other villagers started using the "bridge" pretty quickly after that demonstration. Thereafter, we proceeded to set up the study site, which was a half-hour walk from the bridge, and an on-site "lab" (a screen tent where we could band birds and record data out of the rain). It was during this shakedown period that we all learned a lesson on how to care for one's feet in a place where water often gets into your rubber boots.

We had all previously done research in rainforest, but not where you have to cross streams above your boot tops all the time. For the first few days, we simply ignored this problem and sloshed on. By the fourth day of this (the nineteenth), the skin on the bottom of my feet just sluffed off. Same thing for Sergio. Since we couldn't walk much in that condition, I headed off to San Andrés to get gauze, tape, powder, antifungal medicine, and antibiotic so that we could wrap them up. This treatment helped a lot, and I was able to get back out into the field the next day. By that time, the same thing had happened to everyone else's feet. Sergio and I were the only ones working for a couple of days after that. Turns out it's pretty simple to avoid. Whenever water gets over your boot tops, stop, dump it out, wring out your socks, and continue to march. Then every evening, treat your feet with powder and antifungal medicine. What you can't do is keep walking with water in the boots. Live and learn.

I should mention an additional piece of information that I learned about working with rural people under primitive conditions for prolonged periods, in Mexico and many other parts of the world. Often writers lacking this experience try to imagine what "it must have been like." In their minds, one of the obvious problems would be the stench of unwashed bodies in close proximity. My personal experience has been that *campesinos* and other indigenous folk know quite well how to keep themselves presentable and inoffensive regardless of the environmental challenges and, indeed, take pride in their ability to do so. It took our crew a while to learn how they were able to do this, but we eventually hired a laundress to rinse out our clothes after each day's slog. Continual rain meant she had to dry them over a fire, but woodsmoke is a better alternative to smelling like the north end of a southbound goat, which was the inevitable result of wearing the same unwashed gear every day.

Figure 17-3. John Howe with Agostín and Pascual, *secretario* and *presidente*, respectively, of Ejido La Peninsula de Moreno.

We had accomplished all of our objectives by the twenty-fourth, so we headed out, dropping off Sergio at Xalapa, Jacinto at Veracruz, and Mario up the coast a little at La Mancha. Two of our three vehicles were hors de combat, the whale with brake problems and the INIREB jeep with a blown transmission, both of which were left with Señor Escalera, a master mechanic, in San Andrés. He and his family were to become good friends of ours through frequent and prolonged vehicle issues over the next several years.

Kevin and I headed north after dropping off Mario. Fortunately, we left him with the shotgun and shells. We were stopped and searched three times for guns by the army before finally reaching the border at Matamoros, where we had one more little incident. Driving up to the customs booth, I knew we had a problem. Each of our visas that we were going to hand over to the customs inspector said "Entry with vehicle," but we had only one vehicle, the whale having been left at Señor Escalera's. This is the kind of offense, tried often by smugglers, for which they put people in jail. However, as we pulled up to the inspector, another inspector came over to talk with him. As he turned away, I just dropped the money for the toll and the visas on his counter and took off. No screaming or shooting, so we were home free, back in the good ol' USA.

I went down to the Tuxtlas again in late January 1983 to help our students get their various research projects underway at La Peninsula. With me on the drive down from Kingsville were Sergio Barrios, newly enrolled at Texas A&I as my graduate student; his wife, Aurora; and another new graduate student of mine, Coleman Nemerov. Coleman had been one of my top pupils back when I was teaching high school biology in Minneapolis. He contacted me after graduating from Carlton College about the possibility of graduate work, and I said, "Sure."

Two of Mario's students, Isabel Carmona and her husband, Jose Luis Alcantara, along with Jacinto, were already up at the site when we arrived at the trailer at 10:30 a.m. on the morning of January 21. When they returned about an hour after our arrival, they reported the bridge broken, the lab collapsed, and the mist nets a mess. This was no surprise, really, considering that it had been three months since our last work there. After a brief lunch, we went out to fix the bridge, clean the nets, and get the lab ready to function. The next day, leaving the crew to continue site cleanup and organization, I drove to Catemaco to meet John Howe, a photojournalist who had arranged to do an article on our Tuxtlas work for the magazine *Defenders*.

The fundamental purpose of our La Peninsula project, as laid out in our proposal to WWF, was to study species composition, dispersal, and survival of the migrant and resident bird communities of forest and various types of second growth. Mist-netting, banding, and recapture supplemented by observational notes in the various habitats over a period of years would give us a good handle on most of these objectives, but not the key one—namely, differences in survival (fitness) between forest and second-growth inhabitants of the same species. Radio-tracking would have to be used to provide this information, and for that purpose, there was really only one migrant species large enough and common enough: the Wood Thrush.

Mario's students were focusing on individual resident species for their thesis work, while my students, Sergio and Coleman, were to study various aspects of mixed-species flock movement and composition. The Wood Thrush work, I decided, I would have to do myself due to the likely logistical and field challenges. Fortunately, due to a marvelous piece of good fortune, I found a student who could work with Mario and me as a full partner in this difficult enterprise.

The way this happened was as follows. On the twenty-eighth, I drove into San Andrés to drop off Jacinto, Isabel, and Jose Luis at the bus station for their trip back to Xalapa. After leaving them, I went to the *larga distancia* office to try to call Bonnie, where I ran into Dick Vogt, famed herpetologist from UNAM's Instituto de Biología, who was doing some fieldwork at the station. We chatted a bit, and I invited him to come out and visit us at our study site. Then I headed back to La Peninsula.

Three days later, and much to my surprise, Dick showed up at the netting site (demonstrating some significant tracking skills; it was not an easy place to find) along with two undergrad students of Dwain's from Minnesota who were staying at Playa, John Gerwin and Kevin Winker. Kevin was then, and is now, a giant (six feet five) goth mesomorph—physical attributes that I later found were matched by his intellect and mental toughness. Back in the States, we corresponded, and I eventually invited him to study Wood Thrush survivorship in Tuxtla forest and disturbed habitats for his master's degree, an offer he accepted.

I continued working on developing and practicing the techniques for transmitter attachment and following tagged birds for the next few days until our departure for Kingsville on February 5, 1983.

Kevin Winker came on as a new graduate student in the fall as part of the Bell Museum graduate program at Minnesota under Warner's supervision, and on September 29, 1983, we took off for the Tuxtlas to get him started on the Wood Thrush project. I knew that this effort was far too challenging for a beginning master's student; the logistics alone were dreadful, let alone the field and technical skills required to address the central questions of Wood Thrush behavior and survivorship in forest and marginal second-growth habitats in the Tuxtlas. Ideally, master's projects are for training; one does not expect a candidate to address big questions like those for a PhD. However, at Texas A&I, I did not have academic access to PhD candidates. A&I offered no PhD at that time, and besides, I was only a lowly postdoc myself. So my thinking was that if I could find a person with the courage, stamina, and intellect to take on the Wood Thrush work as a master's candidate, Mario and I would provide field assistants for him whenever possible and would do a lot of the project design, setup, and radio-tracking work ourselves. Kevin W. turned out to be that person—in spades.

Accordingly, back in January, as I have described above, I had worked with graduate students to set up the study site and living arrangements at La Peninsula so that several projects could be done there simultaneously, with all of the students pitching in on the mist-netting, banding, and recording that all would require. I also had done some transmitter placement on Wood Thrushes followed by some tracking on foot to test the feasibility. This September trip was just to familiarize Kevin W. with the living arrangements, the people of the La Peninsula community (who would do his washing and provide fresh tortillas), the study site, and the netting and banding routine. Mario sent Jacinto down to help out with this. I stayed for only a few days, then headed back up to Kingsville on October 5.

On December 3, 1983, I headed back down to Mexico. My first task was to attend the Segundo Congreso Internacional de la Sociedad de Ornitología Neotropical. The first congress for this new organization had been held in Buenos Aires in 1979 (as described in chapter 15). The second would be in Xalapa because Mario was the president. I gave a paper there on migrant use of forest and got into a public spat with Chan Robbins concerning the importance of forest to migrants; Chan's paper was on migrant use of second-growth and marginal habitats in Belize. After the congress, Kevin W. and I headed down to La Peninsula, where I planned to spend a couple of weeks teaching him how to attach radio transmitters to Wood Thrushes and then follow them around with antenna and receiver.

The logistics of doing a field project in the Tuxtlas were difficult and the working conditions miserable. Rainfall averages over twelve feet a year, and the roads generally are unpaved and in terrible shape, or at least they were during our work there in the 1970s and '80s. There is a relatively dry season, which usually starts in mid- to late February and lasts into May, but if you are going to work on wintering migrants, which arrive in October and leave in April or so, you have to be prepared to spend a major portion of your time working in the rain. It's a fair question to ask why we choose that area. As already mentioned, it was the closest place to the United States with reasonable amounts of remaining rainforest. I have worked in other areas since—in Belize, Guatemala, Honduras, and Costa Rica—where the logistics might not have been quite as terrible, but the Tuxtlas was where we got started and where we did most of our work in the 1970s, '80s, and early '90s. It was accessible, we knew the bureaucracy, we had excellent in-country collaborators, and we had experience. Nevertheless, the Wood Thrush work to figure out the fitness question was by far the toughest problem to tackle. It took us five years to accumulate enough information to figure things out. The main thing required is prodigious stubbornness, and for this, Mario and Kevin Winker turned out to be the perfect (and probably the only) people who could have pulled it off.

It was during this trip that I had a chance to see a bit of Tlaloc's true power up close and personal. I thought I already knew something about the Aztec rain god's dominion, but as it turned out, I had a lot to learn.

I think that perhaps the best way to convey the difficulties encountered is to quote a section of my notes from late 1983, when I was working with Kevin W. to get his project on Wood Thrush radio-tracking underway. We had arrived in heavy rain at the La

Figure 17-4. Tlaloc and propitiates.

Peninsula study site on December 11, and it had rained hard every day since while we were setting up the nets, capturing the Wood Thrushes, fitting them with radio transmitters, and trying to track them with our handheld antenna and receiver. The rope bridge that we had built during our initial visit in February and then rebuilt during each subsequent visit was completely gone—not only the ropes but the trees to which they had been attached and the channel that we had used it to cross. The river had simply swept it all away, carving a new channel in a different spot where no convenient trees were present for rope and platform attachment. What follows is a transcript from my notes from three days during that initial field season for the Wood Thrush project:

17 December 1983——During the night we had a horrendous storm. Beginning at about 0230h, with thunder, lightning, and torrential rain, it continued until near dawn. Then the thunder and lightning stopped and the rain slowed. The soldering iron [for activating the transmitters] was not working although it was well-charged. Kevin fussed with it for a while. I waited to see what the outcome would be. If he couldn't fix it, he would have to go into San Andrés [to the TV repair shop] to get five more transmitters working. After about an hour or so, he figured out the problem and fixed the thing. I proceeded to connect three more transmitters to their batteries. It was decided that Kevin would go into town anyway to get supplies and to look for eyelash glue [the adhesive used to attach transmitters to the birds]. The old stuff [that I had used in January] had broken down due to age and was not working properly. A third reason for his trip was to fix the back window of the truck, which was not working. I left for the study site and Kevin left for San Andrés at about 1000h. My trip in was uneventful. The rain had become a drizzle, and though the river was up, it was not dangerously high. When I arrived at the site, my first action was to look for the three radio-tagged birds. . . . About this time it began to rain again, though not a downpour. I began looking for 165, no signal. I went to his home area at nets 19 and 20, no signal. . . .

While I was up at net 1, the rain began to come down in earnest. At 1200, I returned to the lab and wrote up my notes. The rain continued hard. At 1400 I went out to set nets 17 through 24 regardless of the rain and to run them at half-hour intervals in the hope of catching old 165. This proved to be a vain hope indeed as the rain kept up hard and steady. There was little [bird] movement of any kind. At 1530, I did see a squirrel (*Sciurus deppei*?) apparently too hungry to hole up any longer. . . .

I closed the nets at 1630 and headed back to camp [the trailer]. The river was much higher than it had been in the morning. In fact, it was the highest I had seen it. Stoney tributaries and side meanderings normally dry were plunging cataracts containing water a foot or more in depth. The entire river, which is usually clear, was a mud brown so that the usual obvious shoals and pools merged into an ominous murky flood of plunging waves. Accompanying this sobering site were even more frightening sounds: the continuous plashing of heavy raindrops falling in myriads onto dense, leafy vegetation; the roar of the awakening river; and most frightening of all, the low thunder of huge boulders thumping against one another, attesting eloquently to the awesome force of the river. I skirted some of the more difficult parts of the river by making a steep ascent and descent of the rocky, vegetation-encrusted wall of the river gorge. Finally, I was forced to make a choice between following the river or taking a steep, long, and difficult overland journey to the *ejido* without the benefit of a path. I chose to follow the river. When I descended to the river, I followed our usual path, which leads along normally dry meanderings of the river. These were filled with rushing water. I had no problems until I came to a place where three different meanderings join to form a true branch of the river. This branch is normally about .5 m–1 m deep and not too swift. This evening the water flowed in a swift, unbroken swell over the normally placid pools. I had a pack on and my waders and I was carrying a walking stick. I was wearing a raincoat and a hat. When I crossed two of the arms feeding the branch, I knew I was in trouble. I had a hard time keeping my feet in these, and I still had the main branch to go. I launched myself across the main branch, a distance of some 15 m, with considerable trepidation. Below the crossing point, our usual ford, the gorge narrows and the water shoots through a rocky defile. To lose one's footing was to be washed into this mass of rushing water and boulders, not a cheery site. As I worked my way across the flooded ford, I felt the water pulling at my body with a strength that was surprising to me. Each forward step was accompanied by a scary moment of uncertainty as I lost a contact point and felt myself near to being swept downriver. When I reached the middle, and with still 7 or 8 meters of water to cross, I realized that I could not make it standing. The water was mostly chest deep, and in front of me was a flume of more concentrated force than any I had passed. I edged on a couple of more meters and then, with 5 meters or so remaining, tried to project myself toward the achingly close shore. The water took me with a swiftness that was astonishing to me and washed me directly down the gorge. I had one last chance before entering the rocky, pummeling waters of that section and that was to grab a rock, which I managed to do. With that handhold, I dragged myself to shore, sans hat and walking stick. The path from here parallels the river. It was flooded and more often a part of the river than not, but [had] seldom more than .3 to .5 m of water over it until it crosses a second branch of the river. This branch is narrow, about 5 m, but relatively deep, about 1 m at normal time and considerably deeper at present. To cross this I launched myself out and angled for the shore, where I arrived without too much difficulty. The path from this spot to our cable bridge [across the main channel] was uneventful. When I got to the cable bridge, I had a surprise awaiting me. I had left the return rope attached to the pulley and harness loosely tied so that a person wishing to cross from the other side could pull the harness across to them. The water had risen enough to catch the return rope and work it free from the hold on my side. The harness and gloves, therefore, were dangling over the river halfway across. I climbed up to the pulley and tied myself to it. Not trusting the rope completely, I hand-over-handed my way across the channel. Our usual landing place is a peninsula, but under [the] present conditions, it was an island completely surrounded by the river. I had to swim with the pack and waders to get to the other side. I finally arrived at the trailer, somewhat bedraggled, at 1800 or so. Kevin had just arrived from San Andrés, where he had spent most of the day getting the car fixed. He had been unable to purchase eyelash glue [and] had instead gotten mascara, much to the amusement of the people in the *farmacia*. The rain continued.

18 December 1983—— We determined to go up to the site to at least check on our radio-tagged birds and net a little if at all possible. The [Coxcoapan] river was down somewhat [from yesterday], and we had no trouble in cross-ing. When we got to the site and tried out the receiver, the batteries seemed to be dead or nearly so. I went back down to the trailer to get replacements while Kevin opened nets 17–24. I picked up batteries, the extra receiver, and a machete at the trailer. On my way back, I made a [crude log] bridge across the narrow, deep area [of the river] and trimmed vegetation along the path. When I arrived at the site [at] about 1300, Kevin was out checking nets. I tried the batteries in both receivers, and all seemed to be dead or nearly so. I determined to go into San Andrés [nearest large town, about an hour's drive under dry conditions] during the evening. Kevin came in with a Wood Thrush about 1500, a new bird from net 18. I proceeded to put a transmitter on this bird. This turned out to be a struggle. The bird moved every 2–3 seconds so that the glue [eyelash glue was used because it is nonirritating to the skin] would not bond the transmitter to the cloth on the bird's back. I finally stuffed the bird in a sock and got the thing on. It was released at 1700 back at net 18. The receiver worked well enough to track birds #1130 and #1182. #1130 apparently dropped his radio. It was found near net 34. No sign of feathers, etc., so we think it just fell off. #1182's signal was coming from about 7 m [20 feet or so] up a tree, the same tree from which an ant tanager transmitter sent signals last year. I believe it is a raptor roost and that #1182 is a dead bird. No signal from #165. We returned to camp without mishap; the river was still up but not too bad. Left for San Andrés about 1830 after bathing. When we arrived at Catemaco [a town on the route to San Andrés], we found the streets thronged with people. A great celebration was underway, having to do, I think, with the days of travel before Christ's parents reached Bethlehem. Anyhow, everybody was out walking, eating, and drinking, but few stores were open. We went on to San Andrés and, after some search, found a *farmacia* that was open and had batteries. We bought a bunch. They did not have eyelash glue, though they were quite amused that we wanted some. We stopped at Hotel Catemaco for supper about 2130, then set out on our return to our trailer. The "road" from Coyame [small village] to La Peninsula [our home village] has begun to wash out again, and there were many bad ruts and channels. It was raining quite hard as we worked our way down the track. The path crosses several streams, only two of consequence. The last one before the trailer is the worst. It usually is deep enough to come up to the floorboards. Tonight the smaller of the two was up to the floorboards, so we knew that the other would be very bad. It was. I gunned it across. The water was over the headlights, but we fortunately didn't stall. I don't think it could be much deeper and still be crossable.

19 *Dec*—— It rained through the night and continued through the day. We went up to the site, though the river was high. As it continued to rain fairly hard, we simply checked locations of our 2 radioed birds, #1182 and #1204, and returned. #1182 was still up the tree, and #1204 was still near net 18, where he was released last night. This accomplished we packed up the receiver and other gear and left it at the lab [a tent at the site]. Then went back down the river to the *ejido* [village of La Peninsula] to our storeroom, where we got a coil of rope. We took the rope back to the bad river crossing, the place where I fell [and was nearly swept down river] on the 17th, and put a line across here as a safety line. This was a bit of a struggle. The river was high, and though I didn't have a pack this time [as I had on the seventeenth], it was strong enough to pull me off my feet. I got the line across and made it fast finally. Just as this was accomplished, a man who traps langostinos [marketable river crustaceans] came up to the crossing with his son. I was unsure whether they were heading back to the *ejido* or not, so after exchanging remarks about the execrable weather, Kevin

and I crossed using the rope. The man and his son looked at the river for quite a while. We watched them to see what their intentions were. It soon became clear that they wanted to cross but didn't dare. Kevin went back across and told the boy, who was perhaps 11 or 12, to get on his shoulders. He then went back across, holding the lifeline followed by the father. The boy's name was Pedro. His face was a mask of fear before and during the crossing. He was shivering all over from the wet, cold, and fear. Kevin and I got back to the trailer about 1400, fixed ourselves some fried tortillas and *lentejo* [*sic*] soup, and settled down to wait out the day writing notes and reading. We plan to head back up to the site tomorrow, rain or shine, to check out our radio-tagged birds.

These notes give a pretty good sense of the day-to-day difficulties involved in the Wood Thrush radio tracking study, although I must admit that the seventeenth was one of my worst days in the field ever from the perspective of physical challenges. I left Kevin W. to continue the work on December 21 and returned to Kingsville.

THE SANTA MARTA EXPEDITION

Here There Be Trolls

Monsters, as every traveler knows, frequent roads, bridges, and the like. The more you travel, the greater your chances of encountering them or their myriad figurative equivalents as described in several parts of this volume. It's just a matter of probability, although the country in which the roads are located definitely affects the likelihood of these encounters. Nevertheless, there are some especially horrifying entities whose lairs are found at places well off from the beaten path. The Santa Marta crater famously is such a place, home to an extremely unfriendly race of hairy giants known as *hombres de mono* (monkey men).

Dating at least from Alexander Wetmore's trips in the early 1940s, biologists had done a considerable amount of work in the Tuxtla region. Indeed, despite the difficulties of weather and terrain, most of the region had been thoroughly explored and its biota cataloged by the time we began our studies there in the 1970s. There was one large chunk, however, of which this was not true—the crater of the Santa Marta volcano. The rim of the crater represents the highest point in the Tuxtla range at 5,600 feet, while the crater itself is more than three miles wide and half a mile deep, surrounded by nearly perpendicular walls on all but the northeast corner, where an ancient blowout breaches the ramparts, allowing the outflow of the Suchiapan River.

Since our first days at Playa Escondida back in 1973, we had talked about mounting an expedition to explore the Santa Marta crater, the sole remaining bastion of the ancient gods of the Tuxtla forest. Locals warned us against the attempt. One in particular from La Peninsula, Modesto Balverde, had heard of our interest and come to warn us. He told us a rather confused tale (he was a little drunk) about a tribe of great, tall men known as *hombres de mono*—"dos metros en altura con pelo en todos lados" (two meters in height [about six foot six] with hair all over). These men were not Indians. They lived in canyons thousands of feet in depth (the crater) and served as fierce guardians of

Figure 18-1. Looking east across the Santa Marta crater from the air. The ridge running from top right to center was that taken by Mario and me to enter.

the remote fastnesses of the Tuxtlas against violation. We gave him a drink, laughed, and said we weren't worried about *hombres de mono*, especially since we had one of our own (Kevin W.), although he was a bit lacking in the hair department. Modesto did not think we were funny. He muttered a bit about gringos (despite Mario's Mexican bonafides) and wandered home.

By March 1985, we had finally been able to line up the time, money, and people necessary to make the attempt. The funds came mostly from the WWF. Our principal justification was to search for possible remnant populations of endangered species previously known to have occupied the Tuxtla rainforest, like the jaguar, tapir, and Harpy Eagle.

There was a migratory bird aspect to the trip as well. A criticism of our migrant work demonstrating their use of undisturbed forest interior (by Jim Karr, a MacArthur disciple, and others) was that, in fact, the forest was not undisturbed. This point seemed a ridiculous quibble to us, because what difference would a path, or a tree cut here and there, or an open field a hundred yards away be likely to have on the forest interior bird community? Nevertheless, the truth was that until you had really been to a completely undisturbed forest and sampled its avifauna, you could not unequivocally state that migrants used such habitats, so sampling of the migrant community inside the crater provided an additional justification for the expedition.

We set off from Catemaco on March 3. In addition to Mario and me, there were seven others, each with a role to play: Marcelo Aranda, mammalogist and expert tracker (he had written a book on the subject); Fernando Gonzalez, INIREB research ornithologist and expert on guans; Eleuterio Gongora, herpetologist; Eduardo Iñigo, sharp-eyed raptor specialist; John Howe, photojournalist (he wrote an article on the expedition for the magazine *Margins*); Kevin Winker, my graduate student, migratory bird specialist, and gringo *hombre de mono*; and Steve Stucker, another of my graduate students hailing from Minnesota who would help Kevin W. in netting, banding, and observing the forest bird community.

From Catemaco we drove around to the city of Acayucan on the southern border of the Tuxtlas, from there continuing on another twenty miles east on the Coatzocoalcos highway to the town of Jaltipan, where we left the main road, weaving our way northward on dirt tracks through Soteapan, Totohuapan, Benigno Mendoza, Venustiano Carranza, and finally, Magallanes, which was as far as the trucks could go. *Ejidotarios*

Figure 18-2. The Santa Marta party on entry. *Bottom row:* Steve Stucker, Mario Ramos, and Fernando Gonzalez; *top row:* Kevin Winker, Eleuterio Gongora, Eduardo Iñigo, Marcelo Aranda, and John Howe.

put us up in their schoolhouse for the night, and the next morning (March 4), using rented horses, we traveled westward to Guadalupe Victoria, a small *ejido* (twenty-two households) perched on the northeastern shoulder of Volcán Santa Marta where we unloaded our gear and supplies and, on the advice of our guide, Alejandro Peña, regretfully sent the horses back. From here to our destination on the rim, the way was too steep and the arroyos impassable for any pack animal other than humans. By a little after noon, we had gotten ourselves loaded up with as much as we could readily carry in our backpacks—Kevin's and Steve's weighed nearly a hundred pounds—and set off. We planned to spend the night at a *casita* lean-to built to protect *campesinos* tending their cornfields from rain, located well up the blowout toward the rim on the far (west) side of the Suchiapan River, which flowed out of the crater.

The path leading from the *ejido* down toward the river, passing mostly through cornfields and tall second growth, was not difficult. The river itself, however, presented a bit of a challenge. It was fairly broad at the point where our path crossed, with shallow shoals and channels and a deep central cataract thirty feet across. A large log, perhaps two feet in diameter, was the only way over this obstacle. With no load, you would not consider it much of a hindrance, but with a heavy pack it gave you pause, so we decided that it was a good place to take a break. It was at this moment that a curious thing happened. One of our crew began to argue with Mario that the whole enterprise was ridiculous. Why the hell were we here anyway? The animals inside the crater would be the same ones outside, and we knew what those were already. Why risk our lives and make ourselves miserable? Fair questions, I guess, but ones that each person had to have addressed for themselves long before they were sitting in the early afternoon sun on the bank of the Suchiapan River. I listened to as much as I could stand (about five minutes' worth) and then said, "OK, boys. Time to mosey on," got my pack on, and took off across the log, followed shortly thereafter by everyone else, including our critic.

It took us four tough hours to get up to the *casita* after crossing the log. I had to stop and rest about every ten minutes because of the steepness of the path and the weight of the load, as did the rest of the party except for Kevin W. and Steve. They stopped too, but only to wait. They were stronger and in much better condition than the rest of us and could have made it there in half the time. It was dark by the time we got to the *casita* and set up camp for the night.

The next morning (March 5), we split the party. Eduardo and I set out toward the rim, blazing a trail as we went. Our task was to find a campsite somewhere along the rim that had access to water and a study site in untouched forest—inside the crater, if possible, or, failing that, in the Sihuapan River valley that ran along the western base of the mountains outside the crater. The others returned to Guadalupe Victoria to bring up the remainder of our equipment and supplies.

The first half hour was tough going, cutting our way through dense thickets of vine tangles and saplings. Finally we made our way into mature cloud forest, and after about ten minutes blazing our way through that, we found a well-kept hunters' path, complete with the remains (a pile of feathers and guts) of a recently shot curassow. We kept on following this trail along the rim for another half hour until we reached a fork where the ground was relatively flat, which looked promising for a campsite. Following the right fork of the path for a few minutes, we were able to find a small stream that could provide the water we would need for a prolonged stay, deciding the issue. Cutting our way through dense undergrowth that lined the forest along the edge of the crater, we could see the walls dropping steeply to the forest on the crater floor, two thousand feet below. The view showed there was no way down into the crater anywhere nearby. So we started heading down the outside of the crater toward the Sihuapan. This descent took us a couple of hours of arduous hacking until we got to the river, a broad, raging torrent with the usual throaty growl of tumbling boulders. We could hear chainsaws working most of the day, west of the river, but with the river on one side and the mountain on the other, this particular piece of forest was still pristine, except for the occasional hunter. It would serve well for Kevin and Steve as a netting site, and the two hours spent clambering back and forth each day from camp to netting site would help keep them in shape.

With our missions accomplished, we scrambled back up the mountain to the campsite, marking our way carefully as we went (lots of fer-de-lances), and from there back to meet the others at the *casita*, arriving at about 3:30 p.m. Everyone was there, Kevin and Steve having made it back by 12:30 and the others by 2:00 in the afternoon. I explained the situation to Mario, and we decided to pack up and leave for the new campsite immediately. I then went off to fill my canteen at a stream. By the time I returned, our worried crew member had talked most of the others out of coming. John Howe didn't need any persuading; he was asleep. So Mario, Alejandro (the guide), Kevin, Steve, and I loaded up and headed for the new camp, leaving the rest of the party to spend the remainder of the day and night at the *casita*. Arriving at dusk, we set our tents, fixed our supper, and went to bed.

A *norte* came in at four in the morning of March 6, bringing strong winds, hard rain, and cool temperatures. Our pup tent (Mario and me) leaked. We got up at dawn and set out for the *casita* to bring up the rest of our supplies and crew, then returned and finished setting up the camp (sleeping and food tents, eating shelter, cooking fire-place). With those chores completed, everyone set out to begin the survey work that they came for: Kevin and Steve to set up nets by the Sihuapan, Eduardo (who had the best eye for movement detection and identification in my experience) to hawk-watch at a look-out with a clear view over the forest; Eleuterio to survey for herps; Fernando to look for curassows and guans; and Marcelo to search for mammal tracks at the streamside. Part of our equipment for the trip was remote-trip cameras on which we hoped to record rare forest species. Mario and I went off with John Howe to set out some of those cameras.

After breakfast on the morning of March 7, Mario, John Howe, and I went out to place more cameras and spent most of the day at that. The next day (March 8), Mario, John Howe, and I left on what I thought would be a two-day trip around the rim in search of a way down into the crater. Skys were clear when we set out after two days of cold rain, but by noon it had clouded up and was raining. We made pretty good progress despite the weather and stopped a bit before dusk to establish camp, eat, and write up our notes. Heavy rain came in about 4:00 a.m. the morning of the ninth and soaked us while we were breaking camp. We moved along the southwestern part of the ridge, where clearing by the *ejido* of Santa Marta, a Popoluca settlement, comes right up almost to the edge of the rim. At this point the rain was coming down quite hard and the wind blowing around the peak (the highest point on the crater and in the whole Tuxtla range). Mario called a halt and suggested it was foolish to continue under the miserable con-ditions. I argued that we had already come quite a ways, and who knew whether future weather would be any better. John Howe voted with Mario, and we set off to return to our base camp, arriving in midafternoon. I picked up the remote-trip cameras that we had set along our outward route and down by the Suchiapan and arranged with our guide, Alejandro, to accompany me back down to the river in the morning for an attempt to enter the crater by following the river upstream.

The two of us (Alejandro and I) left the next morning at 6:30 a.m., bringing along four of the remote-trip cameras for placement inside the crater should our attempt at entry succeed. On the trip down, we passed by the *casita* at about 7:15. Alejandro looked inside, and remarked cryptically, "Nada."

Nada? Nada what? Food? Women? *Hombres de mono?* He didn't say. We got down to the river at about 7:40. Along the way, he made a number of dismal comments: "It's going to rain all day," "We probably won't be able to cross the ford to get upriver into the crater because it will be too high from yesterday's rain," "No one goes into the crater," et cetera.

Once down in the river valley, the path splits, with one branch heading across the river and back up to Guadalupe Victoria, while the other heads south along the west side of the river. Since I had had enough whining about the bad juju of crater entry, I told Alejandro to take the left fork home (Guadalupe Victoria) and get us some supplies (beans, rice, sugar).

Standing at the fork and looking back upriver, you can't tell that you are entering a crater. The blowout is huge, perhaps half a mile across, so the riverside view looks like the entry to a broad canyon. I then took the right fork at about 8:00 to follow the hunter's path that led along the river. At 8:20 I came to a tributary of the Suchiapan that flows in from the west. It is here that overgrown cornfield habitat (*acahual*) ended and rainforest began, and it was this ford that Alejandro had warned would be impassable. Fortunately, it was not, so I crossed and continued on upriver, setting one of the camera traps just inside the forest and pointing across the path, baited with cat scent. I continued along the track paralleling the river through occasional rain with some thunder until I ran out of trail at about 11:00 a.m. From that point I continued on cutting and marking trail as I went for about two hours until the rain became torrential, at which point I sat down to wait until it would let up enough to let me proceed. I knew roughly where I was in the crater because opposite me (to the east), I could see the thousand-foot ribbon of white of a waterfall that was visible from Eduardo's hawk lookout up on the west rim near our base camp. When the rain let up, I set another of the remote-trip cameras and decided it was time to figure out how to get back to base camp. The choices were to head back downriver the way I had come in or continue up the ridge that I was perched on at present. It was getting late (about 4:00 p.m.), and I knew going back would take at least four hours—if the ford was even passable, which seemed unlikely considering the afternoon's deluge. So I decided to continue up the ridge that I hoped might lead to the rim path near the base camp. At that point the ridge was not too steep, maybe forty-five degrees. From here, I will quote my notes for March 10 and 11:

> At 1605h I started up the ridge to see if I could reach the main ridge [rim] path before nightfall. The ridge was not too steep (45–60 [degrees]) for the first part of the way. Then it began to get much steeper. At 1700h I began to hit steep areas (80–90 [degrees]), where the only way up was by grasping roots and branches. These steep areas would run 10–20 m, then even out to thick vine tangles and *huipia* [thorny-leaved, pineapple-like plant] at about 70 [degrees] for 20 m or so, then steep up again. [It was at about this point

while using my machete as a climbing aid (which you should never do) that my left hand slipped down the handle and blade, cutting a deep slice through the web between the thumb and forefinger of my left hand.] At about 1815h, as it was getting dark, I reached what I though was the top—a broad, relatively flat area. It turned out to be a temporary break in the ridge—maybe halfway up. The rim of the crater stretched what seemed to be vertically above me—but it did have some vegetation on it. I started to climb again, but by 1845h it was too dark to see, and the place where I was at had a slope of 70 [degrees]. Above me was one of the 90 [degree] climbs, so I lay down for the night. I was soaking wet and cold. I wrapped myself up in my poncho as best I could and shivered for the next 13 hours. Sleep was impossible. I spent most of my time trying to keep from sliding off the ridge and minimizing the amount of my soaked self exposed to the wind.

11 *Mar*—Fortunately, it did not rain during the night. At 0630h I started climbing again. The weather was clear and cool. I reached the main [rim] path after about a 40 m vertical climb using scraggly bushes as hand holds (damn stuff would grab the pack on my back and try to pull me backward).

We all continued survey work for the rest of the eleventh. Eduardo spoke with a hunter who said that jaguars still lived in the crater, coming out at night to eat cattle. He said that he had not seen tapir tracks for at least five years and had never seen Harpy Eagles (based on pictures in the guidebook). Mario and I started planning for the third assault on crater entry on the twelfth. The first, of course, was aborted due to weather, and the second was my futile attempt to follow the river in. For the third we planned to follow the hunter's trail around the rim to the south side of the crater, where there appeared to be a ridge leading down into it at an angle that did not look to be too steep.

First, however, we had to deal with our disenchanted crew member whose attitude had not improved. We had brought him because we thought that we knew him well, professionally. I, in fact, had encouraged him to apply to A&I as my graduate student. We had wanted his field skills and extensive experience. However, during our week in the field, he had not done any surveys and had proven to be a difficult companion, hard on the group morale. So on the twelfth, Mario asked him to leave and go back to Mexico City, a fairly drastic course of action, since it required him to perform a two-day hike on his own to the nearest bus route, but it had to be done.

Why had it happened? My opinion is that he was unbalanced by the immense indifference of the wilderness: wendigo, jengi, or whatever you want to call it. He was a modern man—intelligent, well trained, with a lovely family in Mexico City. He had a good job and people who cared about him, and he suddenly finds himself on the edge of the wild, where jaguars are not behind bars and fer-de-lances are not in cages. This sensation is not an intellectual position; you can't even tell how it comes or what triggers it. For our colleague, I think, it was the log bridge across the Suchiapan cataract encountered during our initial hike to the rim. You look at it and imagine the possibilities (most are bad), and in your mind you say, "This is pointless. What am I doing here? Why is any of us here? We could die."

If this interpretation is correct, why didn't it have a similar effect on the others? I don't know. Perhaps it did. But a difference was that they had far more experience working in remote places. Also, in my view, most field biologists *want* this kind of experience; they feel a deep sense of awe and privilege when they see a piece of the authentic universe—not Gaia, she's too motherly and civilized; rather, something raw and untouched. I think that was the motivation for many of the people with whom I have worked over the years. They don't enjoy being cold, wet, hungry, exhausted, or in fear of their lives any more than a normal person. It's just that seeing new things in new places makes it all worthwhile. The old saw "It's mind over matter; if you don't mind, it doesn't matter" explains something of what drives the field biologist; it's all about what new sights and insights into the world you can find. The rest is just logistics.

The distasteful duty of our team member's dispatch discharged, we set off on the morning of the thirteenth determined to find a way down into the crater. The weather was clear and warm when we set off at about 8:00 a.m., working our way counterclockwise along the rim. We arrived at Santa Marta peak, the high point of the rim and the whole range at 5,643 feet, at about 11:00 a.m. and our old campsite (from our previous attempt with John Howe) at about noon. At a little after 3:00 p.m., we got to what we called Westernmost Peak, which provided us with a spectacular view of the whole southern and western parts of the Santa Marta portions of the range. Descending the peak, we continued along the rim another few hundred yards to a heavily vegetated ridge leading northward down into the crater, which we presumed was the one seen from our vantage point near base camp on the northwest portion of the rim that might provide our crater entry point. It was too late to begin that effort, so we found a ravine with a small stream where we set up camp for the night.

Figure 18-3. Mario and me on the west rim of the Santa Marta crater, March 13, 1975.

The next morning, we set out northward down the ridge at an angle of about forty-five degrees. It was a bit of a mess, but we were able to cut our way through for the quarter mile or so until the vegetation gradually changed from dense scrub to forest as we got farther down into the depths of the crater, reaching the floor at about three thousand feet at noon, when we set up camp along a stream. The feeling beggars description. After years of thinking, fundraising, and planning, we had arrived.

We set up some mist nets and made observations in the vicinity of the camp for the remainder of the day. On the morning of the fifteenth, I left Mario to continue observations and tend the mist nets while I tried to follow the river northward along the crater floor to the ridge and path that I had tried to use as an entry into the crater from the north five days earlier. The going was fairly easy, being open *selva* for the most part, except where small ridges and streams fed into the main basin. By 2:00 p.m., I still had not reached the large ridge coming into the crater from the west where I had spent the night on the wall, so I began to hurry. I had been blazing trail as I went, but feeling that I was short of time—and since this portion of forest on the crater floor was fairly flat and open, the canopy being a hundred feet or more above me—I stopped blazing and ran whenever I could. At some point in my hurry, I lost my pouch, which had my map, notes, and flashlight. Finally, at about 4:00 p.m., I discovered why the crater had remained inviolate. On the south side of the large ridge where I spent the night was a major cataract coming off the western wall. This tributary had carved a gorge over a hundred feet in depth and forty or fifty feet across where it emptied into the main branch of the Suchiapan, which at this part of the crater was a chasm of comparable depth, but much wider. Also, I could see a tributary emptying into the Suchiapan from the east side of the crater through a defile of similar depth and width. I was probably two hundred yards south of the trail I had cut following the river in from the north, but without wings, there was no way for me, or any other human, to reach it. In essence, the biological riches of the crater floor were strongly guarded by their own moat and ramparts.

Such poetic thoughts eluded me at the time, as I found myself a long way from camp in failing light, with the first portion of the return lacking a blazed trail. I had figured that the river channel would be my guide. Thus I managed to get lost for a good half hour before finally locating the blazed part of the trail. With dark coming fast, I was going much faster than the track warranted—slipping, falling, rolling in *el chocho*, and losing another machete before finally stumbling into camp just before dark.

Early on the morning of the sixteenth, we broke camp and headed up the southmost ridge out of the crater. The weather was cool and cloudy, good for walking as we worked our way back clockwise along the rim north and west. At about 11:00 a.m., just west of Santa Marta peak, we ran into three songbird hunters from Ejido Santa Marta, a Popoluca village on the southwest shoulder of the volcano: a man of twenty-five or so carrying a sixteen-gauge shotgun, a seventeen- or eighteen-year-old teenager carrying a .22, and a child of maybe ten. They had their traps with them, which are small, two-tiered cages, the lower portion of which harbors a judas bird while the upper serves as the trap. The main prey is the Slate-colored Solitaire (*Myadestes unicolor*), known as *clarin* locally. Similar in appearance to a catbird, this thrush, common in highland *selva* and cloud forest, has a beautiful, haunting song highly prized in Mexico City, where a good singer could fetch as much as $125—illegally, of course. The hunters got about $17 per bird, they said, being a long way off from the best markets. It is against the law to capture and sell songbirds, but like many such laws, it is honored more in the breach. We decided to accompany them to their village, given their obvious intimate familiarity with the Santa Marta forests and its denizens. We got to the *ejido* at about 3:30 p.m. and stayed talking with the *presidente* until about 4:00. It was early mentioned in casual conversation that the mother of the young men whom we had met in the forest had recently killed their father, presumably for cause. More germane to our interest, we learned that jaguars were around and that tapirs were not uncommon, not being particularly valued for food or anything else. No one had seen Harpy Eagles. While we were there, seven or eight groups came in carrying *clarin* cages—clearly a major business in March for the good folks of Santa Marta. We had heard that they were involved in the marijuana trade as well. On leaving, we asked for a guide back to the main path, the price of which was a topic of more discussion than normal. We then set off, arriving at 6:00 p.m. at the rim trail, where we set up camp for the night.

I have spent a considerable amount of time among rural people in the United States, Middle and South America, and Myanmar, often entirely unaccompanied, and I have never felt more vulnerable or uncomfortable as I did during our brief encounter with the hunters of Santa Marta. I had the distinct impression throughout our visit that our economic potential was being assessed—as well as their ability to realize it. The contrast between the ambiance of their *ejido* on the southwest slope of Santa Marta with the friendly, welcoming, and helpful citizens of Ejido Guadalupe Victoria on the northeast slope could hardly have been greater.

A strong *norte* blew in during the night, bringing hard rain and cold winds that accompanied us on our hike back to base camp, where we arrived just before noon on the seventeenth. Cold, wet, and half starved, having run out of rations the previous day, we were glad to find that nothing bad had happened during our absence but disappointed to discover that the crew had eaten all of our stores. Alejandro had been sent to replenish them but was not due back until the evening. We spent the rest of the day huddled around the fire with Mario regaling me with his fond memories of Bonnie's Thanksgiving dinners: her savory roast turkey, stuffing, gravy, mashed potatoes, and pies! Finally, Alejandro arrived with supplies for dinner, and Kevin W. and Steve came up from their nets along the Sihuapan. Kevin W. had a tape of a *Prairie Home Companion* episode that we played that night, and pretty much every night when we were at base camp. We knew it by heart by the time we left; it was a lot more comforting than Mario's food encomia.

We packed everything up and pulled all the remote-trip cameras on the twentieth, heading back to Guadalupe Victoria, where we hired a couple of horses to take our gear back to Magallanes and the vehicles. The trip went with no problems.

What did we find? First, we found that many of the same species of migrants lived in the undisturbed forests in and around the crater as were found in the less pristine *selva* of Playa and the UNAM station, such as the Magnolia Warbler, Black-throated Green Warbler, Wilson's Warbler, Kentucky Warbler, Louisiana Waterthrush, Summer Tanager, Black-and-white Warbler, Blue-gray Gnatcatcher, Ovenbird, Wood Thrush, and

Figure 18-4. Santa Marta party remnants on exit. *Left to right:* Alejandro Peña (guide from Guadalupe Victoria), Kevin Winker, Mario Ramos, Eduardo Iñigo, Steve Stucker, and Eleuterio Gongora.

Yellow-bellied Flycatcher. A few were rare or absent—notably, lowland species like the Hooded Warbler and Worm-eating Warbler. At 3,000 feet, even the floor of the crater is at a much higher elevation than Playa, and some species of migrants do not frequent forest above 1,500 feet at this northern latitude.

Second and most importantly, the Santa Marta expedition confirmed that the native ecosystems of the Tuxtla massif were in extremis. No longer protected by Tlaloc (rain god) and Xiuhtecuhtli (fire god), desperately poor people were using the twentieth-century magic of the gasoline engine to subdue the ancient shibboleths and strip every acre of the range to the bone. We neither heard nor saw any evidence of entry by *Homo sapiens* other than ourselves inside the crater: no machete cuts, lean-tos, gut piles, stumps, hat brims, sardine tins, clothing shards, Bimbo wrappers, flip-flop bits, camp-fire leavings, shotgun casings, or Coke bottles. Not so on the rim, where the sound of chainsaws, especially to the west, was a constant reminder of the battle underway with its siege engines at work.

MISTAKES WERE MADE

You should be ashamed of yourself!

—Anonymous heckler at a talk I gave
on our La Peninsula work to the
San Antonio Audubon Society

Our WWF project was designed, in part, to assess the threat of forest destruction, not only as it applied to the migrant and resident avifauna, but also as it pertained to the biodiversity of the region. In addition, we were to work with the people of Ejido La Peninsula and the professionals at INIREB to come up with solutions that would improve the economic situation of the *ejidotarios* while protecting the wilderness. This last part was the rub. Despite the fact that Mario and I were the PIs, ultimately responsible for using the WWF money appropriately to achieve the objectives stated in our proposal, neither of us had any experience in "eco-friendly capacity building." Our partners at INIREB did, but as we were to find out (to our lasting shame), we had no real power over them to ensure that promises made to our *ejidotario* cooperators were kept.

I have explained our initial contacts with the *ejido* back in September 1982, when we negotiated the use of a portion of their remaining forested lands for a multiyear study of *selva* avifauna in general and the Wood Thrush in particular. The goals were to study the differences in species composition of primary forest and various types of second growth and to determine whether there were differences in survival for wintering migrant Wood Thrushes in forest as opposed to second growth.

In return for the use of their land, we agreed to provide $200 a year to pay for supplies for the school, to transport folks and their goods (including a full-grown pig once, to our lasting regret) to and from Catemaco when we were going, and to help in other ways with special needs when we could afford it. For example, when we learned that seven children in the *ejido* had died of pertussis (whooping cough) the previous winter, we arranged to have the kids vaccinated each year. Similarly, when we found out that two men had died of snakebite during planting, we made sure that antivenin was on hand in a cabinet at the school at all times. These arrangements worked well during the initial parts of the study. There were *ejidotarios* who weren't happy to have us wandering around their woods, but they were in the minority, and the leaders—Agostín, Pascual, and their families—were personal friends.

This relationship changed as a result of the shift in focus of WWF grants. Previously, as I have mentioned, Tom Lovejoy was happy to support projects like ours that were purely investigative—that is, we were simply trying to find out how tropical ecosystems work. The concept was that conservation aspects would be built on this heightened understanding. However, after Tom's departure, WWF people let us know that projects that were pure science would no longer receive funding. Instead, priority would be given to those that incorporated both conservation and eco-development aspects. In other words, we could continue our basic science work only if it was tied to specific economic projects in cooperation with the people who lived in our study area. This part of the project was initiated in October 1983. At that time, I spoke with Agostín and Pascual at some length about how we might work with the *ejido* to improve their economy. Right from the start, they made clear that there was only one venture in which they were interested—namely, the construction of a bridge across the Coxcoapan. Now, had they simply wished for a passenger bridge, a modification of our existing zip line to increase ease of access and safety could have been made for a relatively small sum. However, that was not what they wanted. What they wished for was a bridge large enough for the transport of goods—specifically, logs—so that they could cut the three thousand acres of *ejido* land that remained in *selva* and also, of course, so that they could readily transport produce and livestock to market in all weathers. And it is here that the sense of betrayal begins for all parties. Because it is here that the goals of the granting agency and the *ejido* part ways. The WWF wanted to work with local people to preserve the forest; the people wanted to improve their standard of living as fast as possible, which meant cutting the forest. The sad truth is that there is no way to overcome this impasse. After forty years of experience, I now know that. The only way it works is to simply tell the local people that they cannot cut the forest and then work with them to find other ways of making a living.

To the credit of all parties involved, I don't think that any of us understood the situation clearly at the time, except maybe some of the villagers. It didn't take a genius to figure out that without the bridge, the *ejido* as an agricultural enterprise was doomed. Even

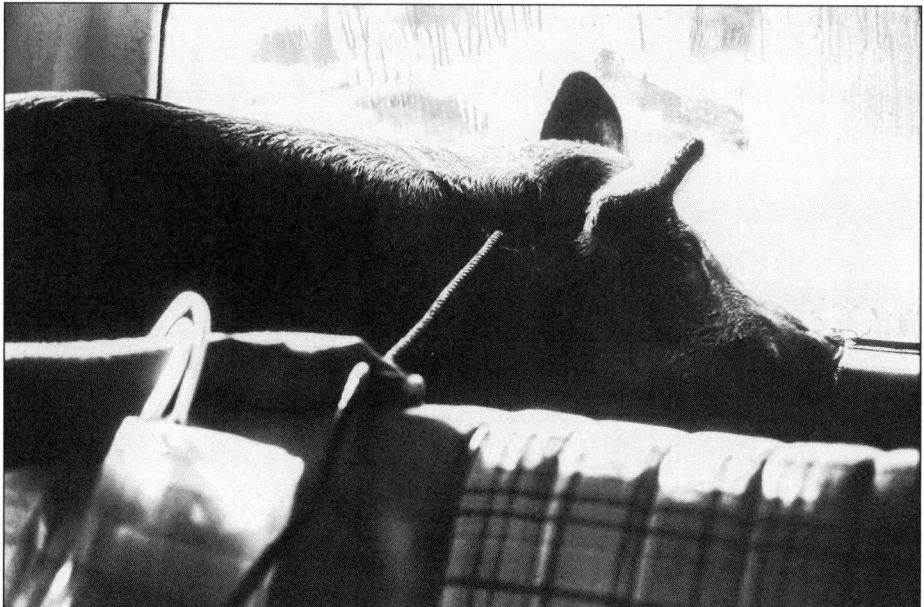

Figure 19-1. The pig in a contemplative moment. The Blazer's atmosphere was never quite the same thereafter.

with the bridge, its ultimate failure would simply be pushed a decade or two into the future. The only difference would be that both the people and the forest would be gone because the Tuxtla slopes do not provide good land for farming or the kind of climate for long-term agricultural success.

When Louisiana State University doctoral student Bob Andrle began his pioneering Tuxtla ornithogeographic studies in the early 1960s, perhaps half of the range remained covered in forest, constituting some of the last acreages available to land-poor peasants through the *ejido* process in the entire state of Veracruz. The lowland, relatively level portions went first, and by 1968, when a group of colonists from villages around San Andrés Tuxtla got together to try to locate property available for *ejido* establishment, the steeply forested slopes of the northwestern Santa Martas, on the wrong side of the Coxcoapan River, were about all that was left. With no other options, these intrepid settlers obtained permission to clear and farm a six-thousand-acre plot, which they named Ejido La Peninsula de Moreno. Pascual, the current *ejido* president, and his parents were among those founding members. He recounted to us something of what it was like to claw one's living from this wilderness, beginning with the fact that it was already occupied by indigenous Popoluca. Like Olmec and Aztec invaders before them, the La

Peninsula *ejidotarios* found these people "problematic" at first, not only because they (the *ejidotarios*) were taking their land away from them, but also because they couldn't explain that they had government permission to take it away because the Popoluca could not speak Spanish. Pascual said that now that most Popoluca were able to speak Spanish, they didn't have so many difficulties with them.

The farming methods used by the new community date back millennia; essentially, it is "slash-and-burn" agriculture, called the *milpa* system in Mexico. This process involves clearing the existing vegetation on the land, usually forest, by first cutting it down and then burning it in place, ensuring that key nutrients go into the soil to nourish the crops that the colonists will plant. The main crop was corn, not because it was the best crop to plant in terms of yield (it wasn't, they informed us). The Tuxtlas are too wet for ideal corn production. Rather, they planted it because they could eat it—that is, they and their families wouldn't starve to death even in a bad year when insufficient corn was grown to be able to market it. In addition to corn, they grew beans, chili, pineapple, bananas, peanuts, manioc, coffee (in the forest understory), papaya, and melons—no tomatoes, potatoes, squash, onions, or cacao because of too much rain. Each farmer worked about 2.5 acres of land, which could be farmed for two to four years, depending on slope and rainfall. Then the land had to be left fallow for several years before a repeat of the slash-and-burn process could be used to replenish the soil's nutrients.

The *milpa* system works fairly well on good soils where land is plentiful so that you can afford to leave cleared land uncultivated for prolonged periods. Also, it helps to be able to grow crops that are best suited for the climate, which is possible only when you have ready access to markets where you can get a good price for your product (so that you can purchase food). It does not work well where land is scarce and markets unreachable or where cleared land is subject to complete denudation caused by landslides— inevitable when steep slopes subject to heavy rainfall are cleared of forest. All of these situations applied to La Peninsula as well as several of the other *ejidos* in the region. The *milpa* system in the Tuxtlas is not a prescription for long-term economic success, and the *ejidotarios* know this. No one chooses to be a subsistence farmer. The people of La Peninsula wanted the same things the rest of us want—a comfortable living and a future for their children better than what they had had.

Conceptually, our partners at INIREB were supposed to be able to come up with schemes to address the economic difficulties of subsistence communities. It is what they were constituted for. So we worked with Mario Fernandez, head of the INIREB agronomy extension section, to develop our proposal to WWF, and in September 1984, we got the money. In early November, Mario and I sat down with Mario F. at the INI-REB headquarters in Xalapa to discuss the economic possibilities for our collaborators at La Peninsula. He said that he would dispatch a team to La Peninsula to obtain basic information on the community—size, soils, crops, families, and so on. Once he had these data, he said he would get training programs initiated for whatever projects the *ejidotarios* might choose. He also said that he thought that he could get cable for the planned bridge for free from the Mexican national power company. We were excited and encouraged by this meeting. We had been worried about our lack of experience in the field of rural economic development, but our meeting with Mario F. had reassured us.

Feeling better about the whole thing, we set off for the Tuxtlas and a meeting with the *ejidotarios*. Mario ran this meeting, which was memorable. He was the best person I have ever seen on his feet in front of any gathering, regardless of size. Personally, I never speak without notes (or better yet, a script) if I can help it. Mario had no need of any such props. He could stand up in front of a neutral or even a hostile assembly, say exactly what needed to be said, and respond with perfect logic and courtesy to the most ridiculous question—or, if things started to get out of hand (not unusual in meetings with people of cultures unfamiliar with Roberts's Rules of Order), with the necessary force of personality to bring people in line.

The result of the meeting was that several *ejidotarios*, led by our friends Agostín and Pascual, agreed to take a chance and devote some of their precious energy and time to whatever projects the INIREB experts might suggest. However, the group made clear that their number-one priority was the bridge—first, last, and always.

As mentioned in chapter 16, I had left the Conner Museum in May 1983 and joined the staff of the CKWRI. Like INIREB, this organization had an agronomy section, and when we got the WWF money, I approached them for assistance with our eco-development work at La Peninsula. Charlie Russell, who had a PhD from the University of Georgia in tropical forestry, agreed to help. He was an expert on intercropping (the use of the same field for more than one crop) and alternate-year cropping (the use of nitrogen-fixing crops in alternate years for nutrient replenishment). I also recruited an agronomy graduate student, Ben Robles, and got him started with investigating the possibilities of terracing to improve the yield and longevity of the farm fields at La Peninsula.

Both Mario and I were busy guys with families and real jobs where we were expected to show up for work in our offices most days unless arrangements had been made ahead of time for us to be gone for a month or so in the field. Thus several months passed after our meeting with Mario F. at INIREB before we were able to get back to La Peninsula. There we found that while our graduate students were working diligently on their individual studies on migrant and resident bird ecology, nothing whatsoever had been done by the INIREB team—no visits, no evaluations, nothing. The only eco-development project underway was that of my student Ben Robles, who had managed to team up with an *ejidotario* to perform the backbreaking work of constructing and planting a series of terraces. Our partners in the *ejido* were not happy. When we left, Mario was determined to find out what was happening with our INIREB team.

It took a while, but we finally got a date and time for a conference with them in Catemaco, set for September 15, 1985, ten months after Mario F. had promised immediate action. This meeting was one of the most depressing of my life. First of all, we discovered that while three of the team members had been to La Peninsula, the team leaders—Armando, the economist, and Victor, the sociologist—had not, and they did all the talking. Secondly, they (Armando and Victor) laid out a socialist agenda entirely at odds with the reality at La Peninsula. They told us that nothing was going to happen on their end until they could see that every *ejidotario* was on board with a commune approach, where all members of the community agreed to take on the projects assigned to them. Mario explained to them that this was not what Mario F. had promised. He had said that after visits by his team had assessed the situation at the *ejido*, a list of potential projects would be prepared, which would then be presented to all members of the *ejido*. Then those members who wished to participate could select the project they thought might work for them. Then they would be taken by the team to a model village in the Tuxtlas, La Vigia, where they could see the project in action. If they thought they could do it, then a team member would be delegated to help them get the project underway, with Mario and me providing the start-up funds.

Armando and Victor said that they did not know what Mario F. had promised, but what we described was not how things were supposed to work. In fact, what we proposed was unprofessional. Without everyone in the *ejido* on board and willing to pull together in a communal effort, nothing would happen. Everyone had to commit to the concept as a whole.

Mario said this was an absurd idea. La Peninsula was not a concept. It was a real village with sixty families—nearly four hundred eating, sleeping, working, breathing people. Where were the funds supposed to come from to create a commune from this community? How were the people of La Peninsula supposed to feed their families while these various projects were getting off the ground? And what if the projects didn't work?

Were Armando and Victor personally going to guarantee that no economic difficulties would arise and, if they did, that they would underwrite the problems? The job of the INIREB team was to get with the program. Figure out what projects might work today, which is what this meeting was for.

Well, there was a great deal of further discussion, back and forth, until the INIREB folk finally told us that there was no way they could help us. In fact, they wanted to know where the money was for us to start projects at their model village, La Vigia. They had told the villagers that we (Mario and I) were going to help them financially, and now here we were reneging. Mario responded that we had never made any such deal. La Vigia was their problem. He said he would sort all of this out with their boss, Mario F., back in Xalapa.

After this disastrous encounter, we headed off to La Peninsula to talk with Agostín and Pascual. They were extremely unhappy with us. They said that no progress had been made on the bridge we had promised, nor had they seen any funds to initiate the projects promised by the INIREB team. This discussion ended with them telling us that our research group was no longer welcome on *ejido* property. That was enough for one day, so Mario and I left to figure out what we had to do.

During some difficult meetings with our friends at the *ejido* over the next couple of days, we found out that their unhappiness with us stemmed principally from a meeting that they had had with Mario F. at INIREB headquarters in Xalapa back in late 1984. At that meeting, Mario F. had promised that INIREB would build a bridge for them by March 1985 (to be paid for by Mario and me with WWF money), that we (Mario and me) would pay $450 to the *ejido* for the use of their land in our research, and that an array of eco-development program options would be presented to *ejidotarios*, also by March, which would cost them (the *ejido*) nothing to initiate. After hearing this fanciful list, we were surprised they were still talking to us. Here it was September, six months later, and there was no bridge, no money from us (about the promise of which we knew nothing), and no sight of anyone from INIREB with program options. You can't mend broken dreams, so we just tried to be honest, explaining that there would be no bridge built by us. Mario F. had told Mario and me back in 1984 that one could be built for $1,200, which we agreed to pay for. However, when we were presented with the plans in early 1985, the actual cost was $12,000, and it was for a footbridge, which we could not afford and which would do little to advance eco-development. We further explained that we were not going to pay rental fees for netting birds on forested land useless for agriculture (greater than a forty-five-degree slope) when we were already paying several hundred dollars a year to the *ejido* for school construction and medical supplies. The only thing we could promise was that we would try to push INIREB to present them with doable eco-development options, for which we would foot the bill within reason, as long as they agreed to do the work.

When Mario got back to Xalapa in late September 1985, he managed to light a fire under the INIREB eco-development people to get them down to La Peninsula with some project suggestions: apiaries, freshwater prawn nurseries, pig farms. Only five *ejidotarios* agreed to work with them, and all chose the pig farm option.

By May 1987, our eco-development work with the *ejido* was over. The pig farm had failed. The fancy pigs we purchased as breeding stock proved to be sickly in the Tuxtla environment and the INIREB veterinarian had quit, so no one provided the nutritional and medical care they needed. Also, meat prices were poor when the pigs came to harvest weight, and the market far away. Ben Robles's terrace project also failed. Ben had done an amazing job with it, demonstrating with his own hard work using local tools that terraces could provide long-term productivity on the steep slopes of the Santa Martas—providing they were properly constructed, maintained, mulched, fertilized, and planted with crops suited to the environment. But the *ejidotarios* would have none of it. The only way such a system could be instituted would be for some outside agency to pay the farmers a living wage (for food for themselves and their families) to construct and plant the terraces.

The one thing that Ben's study made clear was that cutting forest to plant crops on slopes greater than thirty degrees, which included most of the *ejido* land, was a recipe for disaster. This scenario became obvious during November 1986. The rains that month were especially heavy, and when we arrived at Catemaco on the eleventh, we found that all of the mountainside *ejidos* were suffering terribly—access roads blocked; entire hillsides washed out killing, inhabitants living below them; crops lost (of course); rivers impassable; and disease (whooping cough and dengue fever, which I contracted on this trip) rampant. We made our way out to La Peninsula, crossing the river that evening in the dark to find most of the town waiting for us; we were the first to make the crossing in a week. Our purpose in making this trip was to pack up all of our field gear to move over to the UNAM biology station, essentially ending our bird work at the *ejido*. INIREB continued to do a few things for another year or so, but nothing came of it. Like the song says, it all came down to smoke and ash.

Sometime later, I was giving a presentation on our Wood Thrush work in Veracruz to the San Antonio Audubon Society wherein I mentioned the eco-development work at La Peninsula coincident with our study. After I finished, I asked for questions. A young woman in the front row stood up and said, "I don't know how you can live with yourself after what you did to those people. Here you stand with your smug little bird talk, and what has happened to them after all of the promises you and your colleagues made and broke? You're here, nice and comfortable, and where are they? Still down there desperately trying to survive. You should be ashamed of yourself!" I made some lame response about the difficulties involved in working with subsistence farmers on land that never should have been cleared, but I can tell you that was a hard moment. Because I knew she was right. Had we stuck to the research, we would have been fine. But when we agreed to do eco-development, we became complicit in the inevitable bad outcome.

Mario was responsible for whatever positive effects came out of our Tuxtla conservation efforts. He ended his career as senior conservation officer for the World Bank and from that position was able to broker the creation of the Tuxtlas Biosphere Reserve, a huge win for habitat preservation in Mexico but likely of little benefit to our friends at La Peninsula.

CHAPTER 20

WOOD THRUSH LIVES

Still I heard the wood thrush sing,
as if no higher civilization could be obtained.

—HENRY DAVID THOREAU, *WALDEN*

It is difficult for those who are not field biologists to understand, but the goals of my work in the Tuxtlas had nothing whatsoever to do with eco-development or conservation. In fact, I have done my best in subsequent projects to steer clear of these value-driven endeavors. I went to the Tuxtlas to delve into "nature's infinite book of secrecy"—specifically, as it related to the ecology and evolution of migratory birds. From that perspective, the La Peninsula work on the Wood Thrush was an unqualified success.

Many people contributed to that outcome, but the prime mover was Kevin Winker. I have alluded to the personal and logistical tribulations of conducting research at La Peninsula (chapter 17). Those were just a sampling. Kevin spent three field seasons on the project from November 1983 to November 1986, often on his own in horrific conditions, some of which he discusses in his book *Moments of Discovery*. I must say, however, that I find his description of our activities on the night of November 11, 1986, somewhat understated. Crossing and recrossing the raging Coxcoapan with heavy loads in the dark was one of the worst experiences in my professional career, and Kevin was weak and ill at the time with diarrhea! In any event, the outcome of his superlative efforts, as presented in his master's thesis for the University of Minnesota, provides the most complete exploration of the wintering behavior of a migratory bird yet produced.

There were three key elements involved in the investigation of Wood Thrush behavior at La Peninsula: mist-netting (capturing, banding, color-marking, attaching the transmitter), observation, and radio-tracking. Using these techniques, Kevin was able to construct a cohesive, data-based picture from his sample of 217 Wood Thrush captures during his three field seasons as summarized in the account below.

Wood Thrushes begin arriving in the rainforest of La Peninsula by mid-October, with those that held territories the previous year returning to their winter homes and new birds fighting to establish their territories in areas left vacant by failed returns. Specimen data documented that territories were defended by members of all sex and age groups, although we did not determine percentages. Recaptures, resightings, and radio-tracking established the fact that these birds remained on their territories, which averaged about an acre in size, from the time of arrival until departure in mid-April, seldom or never moving beyond its boundaries. Territories were established on arrival by vocalizations, usually the *"Bup bup bup"* call (which changed to *"Bip bip bip"* or even screeches as the birds got more excited), although song was used occasionally as spring approached. These calls were given most often in the mornings and evenings or during periods of low light preceding rainfall during the day throughout the season. Intruding Wood Thrushes normally were chased when seen, and if they did not depart immediately, they were displayed to using the same displays as those seen in territory encounters between males on the breeding ground. Roughly half of thrushes captured in the forest were not territory owners. These birds were intruders onto established territories and, like the owners, came from all sex and age classes at unknown percentages. Based on recapture, resighting, and radio-tracking, intruders remained on the study site for periods ranging from minutes after release (occasionally chased by a vigilant owner as it came out of our hands) to a few days (during *nortes*). We called these thrushes "wanderers," defined by movement of more than 150 yards from the point of capture. The number of wanderers captured in the forest declined throughout the season, but members of this class were still being captured in forest on other birds' territories in March. Territory owners seldom carried more than a small amount of fat (in the cleft at the top of the breastbone), whereas wanderers often carried moderate or even heavy fat loads. After the migration period was over (late November), influxes of wanderers onto forest territories generally occurred during *nortes*, cold fronts often accompanied by heavy rain lasting a few days. During these times, small flocks of a few wanderers might be seen on a territory, with the owner foraging nearby and seemingly ignoring them. We believed that these wanderers might have come from territories at higher elevations whose resources became marginal during cold, wet weather. This belief was strengthened by observation of a Wood Thrush captured by Kevin and Steve Stucker during our Santa Marta expedition (see chapter 18). This bird had a territory at the high end of the elevation range in the Tuxtlas for this species, 1,500 feet. During *nortes*, it disappeared (downslope?), only to reappear when the weather ameliorated.

Kevin's construction of the winter life history of the Wood Thrush was splendid, laying bare how the bird lived during this period in fantastic detail. There was just one critical piece missing from this puzzle, which was to find out whether there was a cost to be paid for wandering as opposed to remaining in a territory throughout the winter season. Kevin's work had indicated that there was a steep price indeed; although he had found only two dead wanderers, he had found zero dead sedentary birds. Nevertheless, it was clear that we needed to be able to follow individual wanderers for longer periods to find out about their movements and the dangers faced. Kevin had done his best to track wanderers on the ground, but a movement of a few hundred yards by a wanderer, a matter of a few seconds of flight time, could take a bird over the nearest ridge—well beyond the detection range for a tracker on foot. We had to get up in the air to find out how far birds were going and what was happening to them. We had known this from the beginning, of course, and in our original proposal to WWF, I had included the cost of an ultralight aircraft, about $5,000 at the time. As one of the reviewers (Jim Lynch, I think) pointed out, this request was ridiculous. He commented that even assuming I could get the necessary permits to import the craft and permission from military air control to fly it, I would be likely to kill myself endeavoring to follow birds over the rugged landscape and in the terrible weather of the Tuxtlas. So WWF forced us to cut the ultralight out of the budget, although they left money for flights in. That left us with the alternative of trying to locate and rent a private plane. This quest proved quixotic.

CHAPTER 21

TEPĒYŌLLŌTL STIRS

From down below Poseidon
shook the boundless earth and towering heads of mountains.
The whole world quaked.

—Homer, *The Iliad*, book 20
(translated by George Cowper)

Kevin began fieldwork on wintering Wood Thrushes at La Peninsula in December 1983, but it was not until September 1985 that we had the funds to pay for flights. He had done all of the radio-tracking of birds on foot up to that time. On the eighteenth, Mario and I set out to find a plane, beginning with discussions with our mechanic in San Andrés Tuxtla, Maestro Jorge Escalera. During the previous twelve years, we had spent many hours (and many hundreds of dollars) using his mechanical expertise and cooling our heels in the family eatery, Restaurante Olimpico. The Maestro was a pillar of the San Andrés business community, and when we mentioned that we needed a pilot, he said he knew a guy, Moises Estrada, who had a private plane and who might be willing to work with us. We called him up, and he said he was interested, and while he couldn't fly this week, he would be happy to talk with us the next day about possible future flights. With that appointment arranged, we set off to continue our investigation of other charter possibilities. We had heard that the town of Lerdo de Tejada in the lowlands west of the Tuxtlas had an airfield, so that was our next stop. We discovered that this was a private strip, owned by a sugar baron. There were three single-engine Cessnas parked there, one of which belonged to Moises. No charters. In fact, the only person present was a muscled, scowling guard in a Mötley Crüe T-shirt armed with an M16. From there, we drove ninety miles north to the Veracruz airport, where we found an air taxi service, which would be happy to fly us wherever we wanted to go for $75 an hour. We arranged to take a flight over the Tuxtlas at 7:30 a.m. the following day.

The next morning, at 7:20 a.m. on September 19, 1985, Mario and I were standing on the tarmac awaiting our ride when the ground began to move—and continued to do so for three or four minutes. It was a weird, dizzying sensation.

Tremors are common enough in the Tuxtlas. Indeed, a powerful one threw Dwain, Mario, and Dick out of their beds at Playa Escondida back in August 1973 and opened a large crack in Raul's restaurant's ceiling. Nevertheless, we knew immediately that such a quake in the flat lowlands of the Veracruz coastal plain likely signaled something seriously bad in the mountains to the west. It was not until later that we found out that a quake measuring 8.0 on the Richter scale had hit Mexico City, causing at least five thousand deaths. In the meantime, however, we met our pilot, a septuagenarian with forty-six years of flying experience, and set out for the Tuxtlas. We explained that we wanted to tape a couple of Yagi 5-element antennas to the struts of his nice little Cessna five-seater and then fly out over the Tuxtlas to see what range we could get in attempting to pick up the signal of a transmitter left on the ground at La Peninsula.

There is an adage in flying lore that states, "There are bold pilots and there are old pilots, but there are no old, bold pilots." Ours provided further evidence of its truth. Except for a few days in April and May, the Tuxtlas have a cloud cap, and that day, in the heart of the rainy season, was no exception, although the ceiling was passable at about three thousand feet. Once in the air, our pilot immediately signaled his uneasiness, but we pushed hard to keep him on his flight plan. When we got over Lerdo, to our complete surprise, he landed at the private strip where we had stopped the previous day.

Once on the ground, he said it was too cloudy to fly over the Tuxtlas today and that we had already used up forty-five minutes of airtime. We said that was baloney. We had used thirty minutes of airtime, and if he didn't get us over the Tuxtlas, he wasn't going to be paid anyway. After a few more minutes of argument, he finally agreed that the air time was thirty minutes so far, and he would try to get us over the Tuxtlas. Unfortunately, however, the plane wouldn't start. In the meantime, the guard, who was even less happy with us today than he had been yesterday, had called the cops, who now arrived on the scene, a lieutenant colonel of the *federales* with two uniformed henchmen (one with a drawn .45 and the other with a submachine gun carried at the ready) plus the airstrip manager (from the sugar company). The colonel grilled us for a few minutes before we were able to convince him that we were the hopeless idiots we appeared to be, after which the posse left, leaving us alone with a dead plane and a truculent guard. At this point, the pilot headed into town to see if he could find a mechanic, returning sans mechanic at about 10:30. At 11:00, a guy roared onto the strip in a fancy new Grand Marquis. It turned out that he was a businessman and lawyer here to meet a plane from Mexico City. He told us that the quake had hit the city hard and that all phone lines and TV from the city were down, essentially cutting off communications and sending the rest of the country into a state of panic—nearly everyone in Mexico has loved ones in the city, including Mario and his wife, Isa, of course. Shortly after this arrival, a mechanic of sorts arrived. He started yelling and bossing everyone around, but eventually, after first getting the polarity wrong on the jump, he got the battery firing, and the plane started. By the time we got up in the air, it was 11:30, and the ceiling had dropped quite a bit, down to about two thousand feet. Nevertheless, we convinced our man to keep going, and he headed out along the coast and then inland over Sontecomapan Lagoon and up the Coxcoapan River valley to La Peninsula, where we had left a radio transmitter to see what range we could get from the air. The system worked. We picked up the signal about two miles from La Peninsula, a range that likely would have been five or six miles had we been able to gain any altitude. I continued tracking the transmitter from different directions for the next half hour or so and also took some aerial photos. The pilot asked

Figure 22-1. Tuxtlas from the air in 1985, with a closer look at Dos Amantes Pass than our pilot really liked. An arm of the Yagi antenna can be seen in the lower right corner of the picture.

Mario if I was an American or a Russian because if I was a Russian, I was not allowed to take pictures. Funny guy. Finally, with our mission accomplished, we headed back to the Veracruz airport, arriving at 1:00 p.m.

The day had just begun for us. Our Veracruz flight had demonstrated that flying would work to locate a Wood Thrush from the air, a major accomplishment. However, the logistics of having to use the Veracruz charter were significant, even if we could find a pilot willing to fly into the range. So we headed back to San Andrés for our meeting with Maestro Escalera's friend, Moises Estrada Sosa, a local architect and owner of the nice Cessna four-seater we had seen at the Lerdo strip.

Accordingly, we met at Moises's office at four that afternoon. He appeared to be a perfectly cultured professional, widely read with a variety of interests including biology, archaeology, history, and so forth. Also, based on his library, he was a devout Catholic (useful for a Tuxtlas pilot). He said he would be happy to work with us but wouldn't quote a price. He said we could figure all that out later when it came time to fly.

Leaving Moises's office, we headed to La Peninsula for a difficult meeting with our friends and collaborators in the village, arriving there about 7:00 p.m. They were fed up with empty promises from the INIREB community builders. Mario and I had already decided that groveling and money were our only ways forward, so we came prepared to offer both, which worked for the nonce.

The next morning, we filled the truck with the village leaders and headed toward what was sure to be a contentious meeting with the INIREB folks in Catemaco (described in chapter 20). The "road" from Catemaco to La Peninsula was always bad, but it had gotten much worse due to heavy rains and logging truck traffic. Anyhow, as we were rounding a bend crawling up one of the hillsides, our passengers started yelling "¡Hoyo! ¡Cuidado!" I was driving, and I just said "Yeah, yeah, yeah," which nonsense was followed by a terrible bang. The passengers said they guessed I had found the hole. In truth, *hole* didn't quite describe our predicament, as I found when I opened my door and looked down into a chasm ten feet deep where the road was simply gone. The front bumper somehow had caught the edge of the crater, or the vehicle would have toppled down the hill. Everyone piled out, with much laughter and mocking of my "Yeah, yeah, yeah," which didn't help my concentration. Nevertheless, we eventually figured out that since the rear wheels were still on relatively solid ground, we could maybe jack up the front end and, with everyone pulling on a rope that we had attached, move the front left tire onto the remaining bit of road. This procedure turned out to be a slow process—jacking up, pulling sideways a few inches, dropping off the jack, jacking up, and so on. But in the end, it worked. We escaped again, resuming the trip to Catemaco.

The meeting with INIREB went about as expected. Money was doled out by Mario and me, and more promises were made. Mario and I left Catemaco at about five that afternoon on the long drive to Mario's home in San Cristóbal de las Casas in Chiapas, arriving at about two in the morning. We found everyone awake at Mario's house despite the late hour and, like most of the rest of the people in Mexico at the time, hysterical. The earthquake on the nineteenth had hit Mexico City hard, but no one knew how hard. All communication with the capital had been destroyed, so the worst, naturally, was assumed. I already had a flight booked to the city from the Tuxtla Gutiérrez airport, a two-hour drive from San Cristóbal. Mario had planned to take me there at 7:30 or so that morning, but with the catastrophe, he asked if I would accompany his wife, Isa, in an attempt to reach her parents' apartment downtown in the Distrito Federal—if she could get a ticket. I said that I would be happy to.

We arrived at the airport at about 9:30 to a scene of indescribable chaos. The place was packed with folks trying to reach the city. Somehow, Mario was able to get Isa onto a waiting list for spaces on my flight, which was scheduled to leave in an hour. She was number thirty-five on the list.

Computers were down. Phones were down, so the airline had no way of knowing when the plane would arrive. Eventually it did—an hour late. They read off names on the waiting list, and miraculously, Isa had the last seat. Evidently, a lot of people had seen the mess at the Tuxtla Gutiérrez airport and chickened out. Anyhow, Isa and I hopped on the plane and headed into the unknown.

We flew into Mexico City at about 11:30. The scene from the air was appalling. Smoke was rising from ruins at several parts of the *distrito*. On landing, we learned that thousands were dead and that the neighborhood where Isa's parents lived was one of the hardest hit. Road travel was impossible, but the subway was still working, and there was a stop within two blocks of her parents' apartment building.

I'm sure that the forty-minute ride from the airport to our stop was among the longest and most painful of Isa's life. The tension was indescribable. But here I must interject an occurrence of such oddity as to beggar description. There were few people sharing our car, perhaps four or five, and at one of the stops, a clown entered—a young man dressed in the normal attire of a student but with a painted face. As the doors opened, he strode in and down the aisle. He looked around with stylized, birdlike movements, then shrugged, raised his hands, palms up in obvious perplexity, and then exited the car. Despite the situation, one could not help but laugh. It was like the "knocking at the gate" in Macbeth. Personally, I felt that the clown was doing what he could to help people get through that terrible time.

When we got to our stop, the scene that greeted us was not encouraging. One of the buildings clearly had suffered considerable damage. However, the structure housing Isa's parents looked OK. Even so, we mounted the stairs up to the third floor with increasing dread at what might be found. The relief for Isa when her folks answered her knock, all safe and sound, was immense. She entered the bosom of her family, and I set out on my own little saga.

I got back to the airport at about 3:30 and found that my flight to Matamoros had left. I then asked the airline for anything heading north, and they got me on a flight to Reynosa—not ideal, but at least in the right direction. I took off at seven in the evening and got there about 9:30. I then rented a car and drove back to Kingsville, arriving at about two in the morning, completing one of my more dramatic trips.

PARSING THE "WANDERER" ENIGMA

The floater (wanderer) is an integral
part of any territorial system.

—*Moi*

We finally caught up with Moises (the architect/pilot) in January 1986 to seal the deal so far as his flying for us was concerned. He had been friendly and chatty back in September, but when it came time to negotiate a price for our flights, he got a lot tougher. He gave me a long song and dance about maintenance and airfield expenses, finally saying he'd need $250 an hour. Fortunately, I knew I could get a plane for $75 an hour, albeit a couple of hours farther away than Lerdo, and told him so. We went back and forth for a while and finally settled on a $600 one-time fee for a complete engine check and $100 for each flying hour thereafter. Kevin and I went up with him for the first time on January 16, and I was pleasantly surprised to find that he had considerable skill and courage in negotiating the dangerous crags and weather of the Tuxtlas, an impression confirmed by Kevin on subsequent flights over the next three months in search of lost Wood Thrush transmitter signals. One can only speculate as to how an underemployed architect in backwater San Andrés developed the expertise and experience for such daring, high-quality flying.

Kevin completed his Wood Thrush work in March 1986, surely one of the most outstanding examples of the field biologist's art and science. With a sensational long-term effort under execrable conditions and with the help of several dedicated volunteers, he laid bare the functioning of winter population ecology, not only for the Wood Thrush, but likely for any long-distance migrant known to defend winter territory. There was just one point that his superb investigation had left cloudy—namely, the difference in overwinter mortality rates between sedentary, territorial birds as opposed to their itinerant cousins. His data were suggestive, as is the simple logic of considering the relative dangers faced by a bird living on a familiar piece of ground with safe and protected roosting sites and where it does not have to move more than a few yards to locate sufficient food lasting through the entire winter season as contrasted with the life of a wanderer.

Kevin's master's thesis for the University of Minnesota, "The Wood Thrush (*Catharus mexicanus*) on Its Wintering Ground in Southern Veracruz, Mexico," made both the accomplishments of the study and the need to understand the fate of floaters (or "wanderers") quite clear. Accordingly, Mario and I set out to attempt to address this issue in our fieldwork of January 1987 and 1988. We decided to move the study site from La Peninsula in the precipitous and remote Santa Martas back over to the UNAM biology station on the gentler slopes of Volcán San Martín. We also changed our field procedures. Instead of banding and recording all birds captured, we would band only Wood Thrushes, each of which would also receive a radio transmitter. The transmitters would be activated by simply removing a small magnet instead of soldering. I never mastered the delicate soldering technique, often overheating the battery in the process and thereby shortening its life, at times by as much as a week or two from the expected twenty-one days, a problem obviated by the magnet. Also, we had developed a new attachment technique involving a harness looped around the legs and across the back instead of the laborious, time-consuming, and far less secure method of gluing the transmitter to the bird's back. With these changes, time from discovery in the net until the release of the banded and transmittered bird dropped from hours to minutes. We also changed our netting behavior. Previously, we avoided netting in the rain as much as possible. Now that we had narrowed our focus to Wood Thrushes, we developed procedures that would allow us to net in any weather. We checked nets every few minutes during heavy rainfall and carried an umbrella attached to a staff and a dry towel. When we found a captured bird, we set up the umbrella to protect the bird from the rain and thoroughly dried our hands before touching it. Using these techniques we were able to get a large number of thrushes equipped with long-lasting transmitters that would not drop off (as the glued-on ones often had) in a matter of a few days, allowing us to spend the majority of our field time tracking the birds.

There had been many changes at the biology station since our first visit in February 1973. The road from Catemaco, although still dirt, was kept in much better shape by regular visits of a road-grader. Offices, dorms, labs, and a cafeteria had been built on the property as well as a health clinic in an attempt to pacify the *ejidotarios* from Balzapote and Laguna Escondida. We even had a washer and dryer! At La Peninsula, Agostín's wife, Rosa, did our laundry by hand, setting it out on rocks to dry in the rare sunshine or, more commonly, roasting it over a fire in their hut (smoky underwear still makes me nostalgic). Usually, once you got wet, you stayed wet—more or less for the duration. Starting our day at the station with clean, dry clothes and a hot meal was great, but their real value was in allowing us far more time for fieldwork that had been wasted on logistics.

The topography and habitat mixtures also were quite different. The forest outside the boundaries of the station had been completely cleared for agriculture or pasture. This situation provided an ideal mosaic of *selva* for territorial birds and small patches of young second growth for wanderers. It is important to note, however, that all nets were set in forest, so all thrushes were captured originally in that habitat. The relatively gentle slopes of San Martín were much easier to navigate than the ridges, vertiginous cliffs, and rushing torrents of Santa Marta. Wanderers had to fly a lot farther to get out of tracking range at the station. As a result, we anticipated doing most of the radio-tracking on foot or from the truck, but we needed an aerial component for birds that disappeared completely. Unfortunately, Moises had moved away, so I had to find a new provider. I decided that the industrial towns east of the Tuxtlas might have airfields, so in November 1986, I took off with my student Ben Robles to try to find a replacement for the upcoming season. We went to Acayucan first, terrifying a bank manager with our request regarding possible knowledge of a private plane for rent, no questions asked. He had us escorted out of his office by an armed guard. Then we moved on to Minatitlán, with no luck, and from there to the Coatzacoalcos Airport, where we were finally able to find a charter service that would do our flights for $111 an hour.

As I have said, we tried to do most tracking from the ground. Nellie Tsipoura and Dave Delahanty helped me in 1987 and Mario in 1988. We only attempted to fly when the combination of missing birds and decent weather required it, because the logistics of doing a flight were substantial. There was no phone line at the station, the nearest one being in the village of Sontecomapan, which had a 1950s-style wooden switchboard with dangling phone cables and plug-ins and a hand crank to reach the operator. This *larga distancia* office was open only Monday through Friday during set hours. Dos Amantes had a similar setup. Catemaco and San Andrés had more modern equipment but similar office hours, although you might be able to get a line at a hotel or business establishment if you had a friend. Once a line was located, I had to find the pilot with whom we had negotiated our flights, Capitán Zapata, to see if he would agree to fly on the appointed day and time. He had given me six numbers where he might be found. This setup could be problematic, as this quote from my notes illustrates:

20 Jan 87 (cont.)——We [Nellie Tsipoura and I] left [the station] for Playa [where we were staying] at 1700h to pick up our gear and head to Catemaco. Called Capitán Zapata at 1930h but couldn't get him at any of the numbers given. We decided to go [to Coatzocoalcos Airport] in the morning anyhow.

21 Jan 87——Left at 0530h and arrived at Coatzacoalcos Airport at 0745h. Got ourselves organized and then called Zapata. He never got our message to meet us at the airport at 0800h [evidently], so we started calling numbers. The first was supposedly his house. A lady answered. I asked for Zapata. She hung up. The second was no answer, as were the third and fourth. The fifth was busy for several minutes, and [at] the sixth, a woman answered. She wouldn't tell me anything about Zapata but at last put a man on who asked if we wanted to fly. I said yes. He said when. I said now. He said Zapata would be there right away. And he was. Within 15 min Zapata was there.

The flight that followed, in a beautiful six-seater Cessna to which I had attached our antennae, was well worth the effort, as this account documents:

21 Jan 87 (cont.)——We took off at 0840h. It was foggy near Coatzacoalcos, with broken clouds, but the Tuxtlas was basically socked in. It took us 20 min to fly to the station. We were flying at 5,000' with clouds billowing up to that level. I picked up channel 8 [a wanderer Wood Thrush], ch. 11 [a transmitter at the lab], and finally channel 2 [a wanderer Wood Thrush whose signal was lost the previous day during tracking]. But we were too high to localize the signal. I asked Zapata to fly below the clouds. He said no, the clouds were too low and thick. We circled for 10 min or so, and I asked him again. He said no, that he would fly to Lerdo, put down, and wait until it cleared. But then he went well out over the ocean and gingerly brought the plane down below the clouds and then worked his way to land. The ceiling was at about 1,200'–1,500', which didn't give us much room to spare inland, but [it was] enough so that we could find the bird. I pinpointed ch. 2 as being about 2 km farther north from his last point [of detection] in a patch of young second growth in the middle of a pasture. I couldn't find ch. 3 or 4 or 12. We headed back to Coatzacoalcos, arriving at 1050h.

Obviously, we were lucky to have found the *capitán* on this day. There were other days when things didn't go so well, as described in the following passage from my notes:

16 Jan 87 (cont.)——[I had a transmitter signal disappear while I was tracking a thrush on the morning of the sixteenth, so I decided to try and find the bird from the air along with some others that had vanished.] Having reached this decision, I determined to go to Dos Amates to call Capitán Zapata. Dick Vogt [UNAM herpetologist] wants to fly as well, so he accompanied me. The idea was to arrange for the plane to fly this afternoon. We could not get Zapata at 1400h. Tried again and again from Catemaco. Couldn't get him. Went and picked up the truck at Escalera's [our mechanic in San Andrés]. The brakes still don't work well. Went to get dinner at Siete Brizas [a restaurant in Catemaco]. Just generally wasted time until 1900h, when I finally talked to the captain. It seems that the plane is not functioning. The mechanic will work on it tomorrow morning maybe. I am supposed to call at 0800h. So after a long day of cooling our heels at various restaurants, stores, and *larga distancias*, we headed for home, with me driving the Blazer and Vogt following in the Ford. Just outside Sontecomapan we were pulled over in pitch dark by a truckload of *vigilancias*. I did not see their uniforms at first, but I did see their guns. [Their truck was unmarked.] I thought they were banditos and almost gunned it on through, which would have been bad news. There were six men on the back of an open pickup, each armed with a submachine gun. I would not have gotten far. There have been several robberies along the road [between Catemaco and the station], so these guys were searching for guns. In the course of my body search, they asked to look at my compass and knife. I forgot to ask for the knife back. We were stopped again at the turnoff for Playa. [Got to the *cabañas* at 8:30 p.m.]

Apart from the above, I never had any bandit-related incidents, although crime was high in the Tuxtlas during the Mexican financial crisis of the mid-1980s, mainly drug related. In fact, in November 1985, there was a shootout between drug runners and *federales* near our study site at Bastonal that made the international news. The *federales* had been tipped off to the presence of a large marijuana stash (a common crop for Popolucas), but they were outgunned, and twenty-one of them were killed, some after capture and torture.

Despite these and similar logistical problems, our two months of intensive radio-tracking in January 1987 and 1988 provided the key data for a clear understanding of Wood Thrush winter population ecology, resulting in a paper published in the ornithological journal *The Auk* in the July 1989 issue. I believe this to be the single most important summary of migrant wintering population ecology to date.

The importance of the work lies in the fact that, like the Wood Thrush, many species of North American migrants are known to employ a territorial social system in appropriate habitat during all or part of the wintering period as well as parts of the transient periods, and absent data to the contrary, the key elements of the system for these other species are likely to be similar to those of the Wood Thrush. These elements were predicted based on theoretical work by Jerram Brown and Stephen Fretwell.

Brown suggested that a territorial system of resource use, wherein an individual would use various tactics (such as vocal and/or visual display, attack, and fighting) to reserve critical resources for itself, would depend on the distribution of the resource in space and time and the density of individuals trying to use it. In other words, if the resource is present in large amounts, then it is not worth defending; if it is ephemeral or present only in small amounts, it is not worth defending; if the number of individuals trying to use it is large, then the effort required is not worth taking.

Building on this idea, Fretwell proposed that if the amount of a given critical habitat was limited—that is, if more individuals are present than the habitat can support—and if pieces of it were defended by each colonist in the form of a territory, then the habitat would become entirely occupied by territorial individuals, and all subsequent attempts at entry would be rebuffed, forcing latecomers to occupy suboptimal habitat or continue searching for suitable habitat elsewhere. He titled this model an "ideal despotic distribution." Its use was primarily to explain breeding habitat occupancy, but it seems to fit the use of territoriality during the nonbreeding period as well, with the important distinction being that the failure to obtain a nonbreeding territory results in a lowered probability of survival rather than a decreased likelihood of breeding.

The ideas of Brown and Fretwell appear to predict what we found to be true for Wood Thrushes wintering in the Tuxtla region of southern Veracruz. Birds arrive in suitable habitat (*selva*, in this case) in fall and establish a territory. Older birds return to the territory held the previous year, while young birds must locate and establish one. If they are successful, they become "sedentary birds" and remain on this territory throughout the winter season. If they are not successful, they become "wanderers" and move into less suitable habitat nearby to feed and roost and continue to search in neighboring optimal habitat for an open territory, or they leave the area entirely to search for suitable habitat elsewhere. These wanderers suffer a much higher mortality rate than sedentary birds.

Nevertheless, many will survive the winter season following this strategy—depending, of course, on the amount and suitability of the suboptimal habitat. For the Wood Thrush, "suboptimal" habitat appeared to be young second growth, which apparently had rich patches of food, but resources were less dependable and more dangerous to exploit.

I have mentioned that Kevin Winker found that wanderers had significantly higher fat reserves on average than sedentary birds among his wintering Wood Thrushes in La Peninsula *selva*. This finding seems counterintuitive to many who consider fat reserves as a measure of "condition," where heavy fat equals "good condition," meaning a higher probability of survival, and low fat equals "poor condition," or a lower probability. This perception is incorrect. Fat reserves are not a measure of condition in the sense defined. Rather, fat represents a method for storing energy when more food is taken in than is required to meet current metabolic needs. Its deposition is controlled hormonally through processes that are not well understood. What we do know, however, is that birds and many other types of organisms do not always store as much fat as they can whenever they can; they are genetically programmed to store fat at particular times or under specific environmental conditions. For instance, migratory birds spend a month or two during the postbreeding period eating only as much as is required to meet their daily activity needs. Then a week or two prior to departure on migration, their behavior and physiology change: they eat intensively (hyperphagia), storing the excess energy consumed as subcutaneous fat, with individuals of some species adding as much as 50 percent to their body mass. Once stores are sufficient, they depart. After completing migration and arriving at their destination, their physiology and behavior change, and they no longer take in more food than their daily needs require and no longer store large amounts of fat. For territorial birds in winter, few individuals carry more than light fat loads for several months until the time for departure on northward migration nears. However, wanderers are confronted with a situation not dissimilar to that of a transient. In essence, they are in a state of perpetual migration until they are able to establish a territory, taking advantage of resource patches whenever they can find them and storing excess intake as fat. That, at any rate, is my hypothesis.

Another misconception exposed by Kevin's work is that the density of individuals is an accurate measure of habitat value. In fact, as demonstrated by our Wood Thrush observations, density is often a measure of habitat instability, where large numbers of individuals are found at temporary food concentrations and low numbers where resources are stable for long periods.

January 1988 was the last time that Mario and I worked together in the field. Shortly thereafter, Mario left INIREB, after being offered its directorship, to take up a visiting fellow position with WWF. After a couple of years with them, he took a position as a senior conservation consultant with the World Bank. We often talked of getting back to the Tuxtlas to continue our study of its avifauna, but to my lasting regret, it never worked out.

ROCKET SCIENCE

The Costa Rica Escapade

In my Father's house are many
[green] mansions.

—JOHN 14:2

Dr. George Van Nostrand Powell is the same age as me. However, because of my military adventures and failure to find a job in my field for three years after getting my PhD, he is several years my senior in a professional sense. After completing his doctorate at the University of California, Davis, on mixed-species flocks in the cloud forests of Monteverde, Costa Rica, he immediately had procured a research position with the USFWS at their facility in Patuxent, Maryland, and after a few years there had moved on to another top-flight research job with the National Audubon Society, focusing on the Florida Bay ecosystem at the southern end of the Everglades.

I first met George at the Smithsonian migratory bird conference in Front Royal in October 1977. There he gave a paper on his observations regarding the migrant's role in the mixed-species flocks in Costa Rican highlands. What struck me then was his focus on precisely what the data told him. He clearly expressed his familiarity with the prevailing theory of migrants as interlopers in tropical habitats as well as the concept that their presence in such an iconic, untouched tropical environ would be challenged by indigenous "ecological counterparts." With regard to this theoretical requirement, he simply stated the facts as he had seen them: "Aside from aggregations at army ant swarms, agonistic encounters between residents and migrants are uncommon and appear to be evenly balanced as to outcome. In Costa Rica, I observed 108 interspecific agonistic interactions between flock members, of which only 3 were between migrant and resident. Migrants were dominant in 2 of 3 encounters."

Figure 23-1. Dr. George V. N. Powell birding in the Andes.

This comment is perfect George. He knows exactly what he is supposed to have found. MacArthur's "ecological counterparts" should have been evicting migrants from virgin cloud forest. But they weren't. Just sayin'.

It was at this meeting that I recognized George as the best sort of field biologist. The kind that goes where his data take him, not where the ideas of others lead him. Nevertheless, it was not until some years later, probably at one of the annual national meetings of the AOU, that I had a chance to get to know him a little better, which resulted in a plan to do a collaborative project in the Neotropics on migratory bird habitat use.

Both of us had done extensive fieldwork on migrants in essentially untouched (by humans) tropical forest—he in Costa Rican highland cloud forest and me in Veracruz lowland rainforest—and we both knew that 90 percent of these habitats had been cut to be replaced with pasture, various forms of agriculture, tree crop plantations, coffee, and second-growth scrub of varying age throughout much of Mexico and Central America. We also knew that the destruction of tropical forest was correlated with marked population declines for many species of Nearctic migrants that bred in the forests of temperate and boreal North America. Was the correlation indicative of a cause-effect relationship, with forest-related migrant declines caused by loss of wintering habitat, or simply coincident with some other factor or factors that were the real causes, such as the fragmentation of breeding habitat? I should reiterate, I think, that the vast majority of ornithologists, then and now, believe(d) that the latter explanation was the correct one—namely, that wintering habitat for most, if not all, Nearctic migrants was not a critical conservation issue because they were wanderers during the winter, not tied to any particular habitat or piece of ground. Therefore, changes in breeding habitat must be responsible for migrant declines.

Answering the question of migrant wintering habitat importance and availability was, we knew, going to be complicated because although many songbird migrants might *prefer* untouched forest environments, they *used* many other habitats when primary forest—that is, forest uncut or otherwise altered by human activity—was not available. Despite this issue, we felt that the rapidly developing satellite-driven technologies of pinpointing geographic location using geographic information systems (GIS) and identification of habitat type using remote sensing from space (rocket science!) offered some ways to approach the question.

The US Department of Defense had possessed these technologies for years but had only recently made them more widely available to the general business and scientific communities. In fact, they (our military) hobbled the effectiveness of GIS by scrambling the signal output of the satellites so that a large number of readings (a hundred or so) were required to be averaged for each location in order to be able to know your position within fifty feet. Evidently, they did not want their satellites to be providing site-specific information to potential enemy missiles. Regardless of this handicap (which was removed a few years later after the wall fell), the possibilities offered for our work seemed significant, and we decided to try to locate the funds and collaborators to make it possible.

Although George was now working for National Audubon in Florida, he still had strong contacts from his years with the USFWS at Patuxent, and it was through them, chiefly Mike Erwin and Chan Robbins, that we put together and funded a multiyear project to develop methodologies for assessing and quantifying migratory bird habitats in the Middle American tropics.

The country chosen for initiation of the project was Costa Rica, where George had fifteen years of experience of work with migrants and where our tame (relatively) rocket scientist, Dr. Steve Sader of the University of Maine, already had satellite maps of unsupervised classifications of habitats in northern Costa Rica ready for verification ("ground-truthing").

I was based at the CKWRI in Kingsville, Texas, at the time we began fieldwork, and on the morning of November 9, 1987 (the day before my forty-first birthday), I flew out of the Corpus Christi airport accompanied by my redoubtable field assistant Rafa Flores to meet George and his wife (at the time), Harriet, in the Costa Rican capital of San José.

All of my previous experience of the past decade and a half studying migratory birds in the Neotropics had been spent in Mexico—indeed, one tiny part of that vast and complex nation, the Tuxtla Mountain region of southern Veracruz. Nevertheless, I felt myself quite the expert on "Latin America" based on this limited sample. It therefore came as a shock to me to learn that Costa Rica was no more like Mexico than Spain is like Venezuela or Texas is like anyplace else. Indeed, this experience served as my introduction to the ridiculous notion that countries sharing a common official language somehow share a similar culture.

So let me begin with a brief introduction to the lovely land of Costa Rica. Although technically "discovered" and included in the Spanish Empire in the early 1500s (Columbus visited in 1502), it was always a distant outpost of that vast agglomeration of peoples. Indeed, it was never really a part of any empire, even in pre-Columbian times. The sparse scattering of indigenous tribal settlements of groups such as the Bribri, Boruca, and Diquis were simply too poor and remote to bother with. Costa Rica, along with most of its other Central American brethren (except for Belize, then known as British Honduras, which had belonged to the British since 1798), achieved independence from Spain in 1821. It was a member of the more or less nonfunctional Federal Republic of Central America along with El Salvador, Guatemala, Honduras, Nicaragua, and Chiapas (now the southernmost state of Mexico) until 1838, when it proclaimed itself a sovereign country. As a result, Costa Rica has been allowed to go its own way politically and economically for nearly two centuries, escaping, for the most part, the devastating interference of adventurers, oligarchs, and imperial interests so damaging to its neighbors.

The mid-1980s was an exciting time in Costa Rica. They were a small country (2.6 million people) with a relatively literate population (97 percent could read and write), a stable democracy led by President Óscar Arias, and a liberal government that supported the conservation of its natural resources and welcomed immigrants from the

Figure 23-2. Central America with Costa Rica shown in dark gray.

United States and elsewhere willing to share in the building of a bright future. Although its record of preservation of native habitats was no better—and indeed, with less than 10 percent of its primeval forests remaining, quite a bit worse than some of its neighbors—it had a reputation that was high in the international community. As a result, the spectacular beauties of its mountains and coasts, the vibrant economy (thanks to free-trade zone agreements with the United States and others enacted in 1981), decent roads, and relatively friendly inhabitants, mostly unarmed (no standing army), made the country a magnet for Americans with a few thousand dollars saved up—or loaned by relatives, friends, or investors—hoping to start new lives and enterprises in a beautiful new land free from the suffocating racism, antienvironmentalism, anti-intellectualism, and militancy of the Reagan world.

This heady atmosphere was readily apparent in the place chosen by George for us to stay for our few days in San José while we arranged for the rental vehicle, maps, and government permits necessary for us to do our work. This little inn, the Hotel Petit, was a modest *pension* (private sleeping rooms, shared bathrooms, and comfortable common room) located in a relatively quiet residential part of the city.

The communal area had a couple of sofas and overstuffed recliners and four or five small tables surrounded by straight-backed chairs. The room was not large, so although each group might have its own seating, they were certainly not separated from anyone else sufficiently to provide any degree of privacy in their conversations. An hour or so each day spent in this cozy space was enough to make one well acquainted with the various dreams and schemes of one's fellow guests, especially since all conversations were conducted in good old American English. A sample will give you the idea.

One party present each evening during our stay consisted of two middle-aged farmers in sports jackets and ties from Nebraska engaged in intense conversation with a casually dressed Costa Rican man. The topic was the purchase of a few hundred acres of land currently in indigo. What structures were on the land? Housing? Barns? What equipment was there? How many workers? Wages? Families? What was the market for indigo? What other crops might be raised? And so forth.

Another group was the attractive young couple from Philadelphia, recently married, whose parents had pooled their resources to provide a hundred thousand dollars or so, which they hoped would be adequate for them to purchase and get up and running a small hotel on the Atlantic coast near Limón. Most conversation was between the two of them as they excitedly examined the available real estate potentially suitable for this venture, although they did involve an agent one particular afternoon planning a trip to look over possibilities.

Then there were the three men in shorts and polo shirts cagily discussing "suppliers," "product," and "clients" for an enterprise located at a site on the Nicaraguan border.

Limitless opportunity for those with the vision, capital, guts, and know-how to make it all work. Great stuff.

We concluded our preparations for the field season by midafternoon on November 11 and headed north on the three-hour trip across the mountains to begin the research for which we had come in the region selected for our study.

Our project in Costa Rica had three main questions to address:

1. What species of birds wintered in lowland rainforest?
2. What habitats other than rainforest were these birds found in during the wintering period (October–April)?
3. How much of each of these habitats existed in our study area (the region covered by Steve Sader's map)?

The process of answering these questions had begun back in the States, where Dr. Sader had provided us with a large map, two feet on a side, showing the results of an unsupervised classification of a satellite photo of our study region with blocks of various colors representing the previsitation best guess as to what habitats were present.

Since the methods used were basically the same for the next decade of work in Middle America, I think it best to simply quote below a description of them from our first paper on the topic published in 1992:

> A Landsat [satellite] Thematic Mapper (TM) image taken on 6 February 1986 was processed by a computer-aided technique (unsupervised classification) to generate habitat (land cover) categories. In this technique, the computer searches the TM image and assigns each pixel in the data set to units that "look alike" on the basis of multispectral reflectance characteristics. An image analyst [Dr. Sader, in our case] then assigns each unit to descriptive habitat categories. The analysis of the 6 February image produced 10 habitat categories. The TM scene used for this study covers 2,016 km² [778 square miles] of the Sarapiqui region of Costa Rica extending from the central valley (Meseta Central) across the Continental Divide into the Atlantic lowlands. Included within this area is Braulio Carillo National Park and La Selva Biological Station, a complex of protected natural areas which extends from peaks of 2,600 m [8,530 feet] down to 50 m [164 feet] elevation. In this study we limited our analysis of habitat and migrant use to areas less than 1,000 m [3,281 feet] in elevation of the Atlantic slope. We deliberately elected to analyze an area consisting primarily of wet life zones to avoid the need to distinguish among climax communities. This allowed us to concentrate on differences between primary forest and seral stages created by human land-use practices.
>
> The accuracy of the habitat classification algorithm was tested by visiting random points and comparing actual habitat with TM predicted habitat. Sample locations to be used for quantifying vegetation characteristics were selected within each of the two most abundant habitat types, mature forest areas and open areas dominated by herbaceous plants, using a computer-generated stratified random sample. A small sample of second-growth sites was selected for a more limited analysis of that habitat type. All selected areas were at 5 ha in size. Five points, randomly selected within the site, were used as foci for sample circles

22.5 m [74 feet] in diameter. Within the sample areas, the dbh [diameter at breast height] of all stems greater than 7.6 cm [3 inches] was measured. Along east-west and north-south transects that intersected at the center of each circle, all stems within 1 m [3.3 feet] of the transects and taller than 1.5 m [4.9 feet] were counted. Ground cover and canopy cover along each transect were estimated on 10 sample points. Canopy height was estimated to the nearest 5 m [16.4 feet] by visual inspection.

Two methods were used to quantify migrant presence at the sample sites. A variable circular-plot (VCP) census method was used at up to 12 points 200 m [656 feet] apart at all sites. All migrants detected visually of acoustically with a 50-m [164-foot] radius of the center point were recorded along with their distance from the observation point. Also recorded was the vegetation structure in which the bird was observed (short herbaceous, riparian, second growth, edge or forest remnant tree) and whether or not a mixed-species flock was present. If a mixed-species flock was encountered during a census, the flock was followed for an extended period (up to 2 hr) to ensure that all flock members were detected. Although flock members not individually marked, this continuous monitoring interval made it possible to determine whether migrants were following the flocks.

The second sampling method used a 1-ha [2.5 acres] grid of 13 12-m [39-foot] mist nets. Nets were set up in five rows 25 m [82 feet] apart. Odd rows had three nets each spaced at 50-m [164-foot] intervals; even rows had two nets spaced at 50-m [164-foot] intervals but with a 25-m [82-foot] offset from odd rows. Nets were operated throughout daylight for a total of 19 hours or 250 net hours. Neotropical migrants were banded with U.S. Fish and Wildlife Service bands; all other individuals were marked by clipping an outer tail feather. Field work [in Costa Rica] was carried out during November 1987 and 1988 and February 1989.

Those readers who have slogged their way through the above presentation of what we planned to do likely will recognize the demanding nature of our planned program. Figuring out how much of each of the major habitat types existed in our study area was tough enough, but calculating what migrants were using them was considerably more

Figure 23-3. Costa Rica with study area shown in light gray.

difficult, as I will relate. Nevertheless, the work set the stage for the remarkably successful effort in the Golden-cheeked Warbler investigation (chapter 27).

Our reasons for selecting this particular part of Costa Rica were twofold: first, it was one of the few places in the country with any remaining lowland rainforest, and second, Steve Sader's unsupervised satellite map of major habitats covered it.

The rainforest remained thanks to rugged topography and efforts at protection by private individuals (chiefly, the plant ecologist Leslie Holdridge) and the Costa Rican government via the creation in 1978 of Braulio Carrillo National Park. There was some forest remaining outside the park, but none that had not been "high-graded"—that is, in forestry parlance, having had all trees with commercial value removed. Other major habitats were well represented: pasture, coffee, citrus groves, banana plantations, scrubby second growth, and so forth.

Six square miles or so of the protected rainforest was home to "La Selva," the campus (classrooms, living quarters, labs, administrative and maintenance buildings, and mess hall) and research lands for the Organization for Tropical Studies (OTS), an international consortium of universities, mostly US, offering courses for college credit on various aspects of tropical ecology and a venue for visiting scientists to conduct research.

We had made arrangements with the OTS headquarters at Duke University for our work and potential stay at the station, but the local administrators, Debbie and Dave Clark, seemed to have been surprised by our arrival. Nevertheless, they were able to provide us with rooms for the night and access to the mess hall. For the use of study sites in the forest, we had to work up a proposal for their approval, which we did on the spot; agree to their assignment of a specific location in the station land; and maintain strict adherence to station ordinances, such as no cutting vegetation, no killing animals, no littering, and so on.

Thanks to George's contacts in the country, we did not have to put up in the La Selva dorm for long. On the day after our arrival, we moved out to the gorgeous forest retreat of famed tropical ecologist Dr. Joe Tosi on the banks, usually, of the Sarapiquí River.

Once settled in Dr. Tosi's lovely stilt-house home, we began the jobs for which we had come. Our first chore was to set up a grid of thirteen nets at the rainforest site assigned at La Selva by the Clarks.

We got final approval for our designated plot on the morning of November 13 and headed out to set up our sampling grid. Now, everyone knows that you have to be extra vigilant whenever beginning work in a new place. It takes a bit of experience to learn how to navigate and work safely, whether in the big city or in the Costa Rican rainforest. For instance, my years with the Smithsonian in downtown DC had taught me, among many other things, (1) never to cross any street regardless of the color of the light without checking to make sure some maniac, sleep-deprived congressional staffer late for a vital meeting isn't about to make a right-hand turn at thirty miles an hour across your walkway; (2) never to make eye contact with people acting or dressing strangely or carrying a sign (a significant portion of those whom you might meet on any given day on the Mall); and (3) never to protest to Men in Black no matter what the provocation. One such person, at my passenger (Paul Weldon) taking umbrage and reaching across to honk the horn when the MIB decided to park and get out in front of us in the turn lane outside the Executive Office Building, stormed up to my window screaming "IF YOU WANT TROUBLE, I'LL GIVE YOU TROUBLE!" A cliché, I admit, but effective. I did *not* want trouble, and so expressed myself. Having a bad day, I guess.

Anyhow, you get the idea. Both George and I had had a lot of experience working in tropical forests, but not *this* tropical forest, so we set out through the La Selva undergrowth to put up our nets with some caution. Good thing. I had not gotten far in clearing our first net lane when a coiled bushmaster seemed to materialize on the ground a foot ahead of me. Her perfect camouflage blended so well with the dead leaves, recognizing her for what she was was like an autostereogram experience, with this 3D monster image suddenly jumping out at me from the 2D forest floor. And just in time too, for she was located exactly where my foot would have landed on my next step. Jeepers!

Bushmasters are interesting snakes. They have a fearsome reputation due to their size and venom cocktail of both hemo- and neurotoxins, but radio-tracking studies explain why they seldom cause problems for people, unlike their *Bothrops* cousins, such as the fer-de-lance, which cause a lot of trouble for men and cattle, at least in the Tuxtlas. Bushmasters forage almost solely at night and in forest, searching mostly for mammals like opossums, anteaters, and agoutis. Once they capture and consume such a meal, they usually crawl up under a ground palm or similar protected, secluded place and sleep for two or three weeks until hungry once again. The fact that this one was sleeping where she was probably meant that she had been unsuccessful during the past night's hunt and basically caught out in the open by the morning light.

La Selva had rules regarding the killing of animals, including poisonous snakes, and I respected them for the most part. I mean, why bother killing a snake you encounter along your path? You've already seen it, so what's the point? A perfectly reasonable attitude if you are on your way to someplace else. However, if you are going to be working in the area for some time, it's a different story. Or so I felt at the time. Sure, I had seen her this time, but there were four of us working this site, and who knew whether we would see her next time? So I killed her. She was an impressive animal: six feet long and as thick as Ahnold's arm.

As it turned out, this encounter served as a valuable warning. While this was the only bushmaster discovered during our months of work in the region, their cousin, the fer-de-lance, was quite common. We met with several on our forest study plots.

Another lesson quickly learned that first day was to always dress in hat, gloves, and long-sleeved shirt when clearing a net lane. Rafa and I both learned this simple dress code the hard way, contracting uncomfortable rashes from contacts with poison wood (similar to poison ivy) and getting stung by large solitary ponerine ants—an exquisitely painful experience.

Once the nets were in place, Rafa and I left George and Harriet to "run" them—that is, to check them every hour or so for captures; remove the birds; band them; record fat, molt, sex, and age; and release them. They were to continue this work until dusk or heavy rain, when they were to close up the nets until the day was over or the rain stopped.

While George and Harriet ran the nets at our La Selva site, Rafa and I set out in our rented Suzuki jeep to begin the process of determining the accuracy of the unsupervised classification and locating sites for sampling using mist-netting.

You will have seen the methods described above for habitat identification and quantification, but they may not make much sense to you without some additional clarification, which follows. The satellite essentially takes reflectance data at a given day and time of a place selected on the earth's surface—in our case, a segment of northern Costa Rica whose reflectance data were recorded on February 6, 1986. The idea is that each major habitat type has its own specific set of wavelengths of sunlight reflected back toward space from its surface. Based on data from other places where researchers have visited a large number of different habitats and connected the reflectance data with known habitat types, a set of algorithms is developed that make a prediction of what the various habitats at a new site are supposed to be. These groupings of reflectances are used to make a map of a site in which seven or eight major reflectance groupings are assigned colors, like green for forest, yellow for banana plantation, red for pasture, and so forth. This map is called an "unsupervised" classification because it is created as a "best guess" of what the habitats are based on data from someplace else. So now Rafa and I were headed out on the road to visit specific points on the ground in the La Selva region and see just how accurate the algorithms were. This process is called "ground-truthing." The best way to do this is to have a regular detailed topographic map at a scale of one to twenty thousand or so with a see-through plastic overlay where the various reflectance groupings identified by the computer are outlined. Then with the topo map and a GPS unit, you travel to a given point on the ground and record what habitat is there, taking a picture of it and, at selected points, gathering more detailed information on species diversity and structure of the vegetation, including such information as canopy height, foliage density at various levels, and ground cover.

This we did, visiting one hundred different sites, equaling ten sites for each of the reflectance groupings. Keeping in mind that each reflectance grouping is assigned its own color on the unsupervised classification map of the study area, what we found was as follows:

> Green—Tall forest (trees greater than sixty feet in height)
>
> Red—Banana, palm oil, coconut, papaya, and bamboo plantations; low second growth (under thirty feet); canna indica
>
> Yellow—Taller second growth (thirty to sixty feet), riparian vegetation, cane, low forest, banana plantation, old citrus groves (taller trees), palm plantation
>
> Black—Water, shadow (like a cloud or that cast by a mountain when the sun is low)
>
> Pink—Pasture, young citrus, sun coffee (no tree canopy), guava, yucca, banana, hedgerow
>
> Blue—Pasture, young citrus, sun coffee, banana, yucca, hedgerow

Orange—Couldn't consistently link this color to a specific habitat type

Brown—Same as orange; no consistent habitat type

Pink and blue presented the most interesting situation—a cautionary tale, really. Naturally, pink and blue appeared strikingly different in the unsupervised classification map, but not so much on the ground. In fact, there was an excellent demonstration of the problem. Often it is a bit difficult to know exactly where you are when trying to link a given color to a particular habitat. However, there was one site where this was not true. It was a peninsula of grass half a mile across almost entirely surrounded by a bend in the Río Sucio. Half of this field was pink, and half was blue. That there would be any difference in reflectance seemed impossible, yet the computer said that there was or had been earlier in the previous year, when the satellite took the data. Walking across this field, about the only difference we could find was that some of the grass was head height and some was waist height, so maybe the areas of lower grass were bare ground back in February 1986. Anyhow, that site illustrates a problem for the use of satellite reflectance data. Lots of things can affect reflectance other than the kinds of vegetation covering the ground: clouds, angle of the sun, time of the year, age of the photo relative to when it is checked on the ground, and so forth. Which is why you do the ground-truthing. The idea was to ground-truth the map, then send our findings to Steve Sader so that he could change the algorithms to produce more accurate assessments of the habitats as they existed in real life. Then he would send the revised classification, now known as a supervised classification, to us so that we could check it again. However, you can see the problem immediately from the list of colors and habitats given above. Some habitats that are quite different—say, citrus grove and old second growth—have similar reflectances. No amount of ground-truthing and algorithm massaging is going to change that fact. Nevertheless, being able to identify and quantify forest, second growth and tree crops, and open grassland could give us a sense of what wintering migrants were confronted with in terms of satisfying their habitat needs.

Within days of our arrival and initiation of the fieldwork, we knew that procedures for assessing migrant habitat use developed in Mexican rainforest—namely, setting up grids of mist nets—was not going to work in Costa Rican rainforest. Aside from Wood Thrushes and scattered individuals of a few other species, we just weren't catching migrants.

The problem was that Costa Rican rainforest is not like Veracruz rainforest. Our sites in Mexico were located at the northern edge of tropical rainforest distribution, and as a result, the canopy was fifty feet lower and the forest itself much less diverse in terms of animals and plants. Don't get me wrong: Veracruz rainforest is lovely, but the rainforests within ten degrees of the equator, like that of La Selva, are something else. W. H. Hudson describes such a forest in his novel *Green Mansions*:

> Lying on my back and gazing up, I felt reluctant to rise and renew my ramble. For what a roof was that above my head! Roof I call it, just as the poets in their poverty sometimes describe the infinite ethereal sky by that word; but it was no more roof-like and hindering to the soaring spirit than the higher clouds that float in changing forms and tints, and like the foliage chasten the intolerable noonday beams. How far above me seemed that leafy cloudland into which I gazed. Nature, we know, first taught the architect to produce by long colonnades the illusion of distance; but the light-excluding roof prevents him from getting the same effect above. Here nature is unapproachable with her green, airy canopy, a sun-impregnated cloud; and though the highest may be unreached by the eye, the beams yet filter through, illuming the wide spaces beneath—chamber succeeded by chamber, each with its own special lights and shadows. Far above me, but not nearly so far as it seemed, the tender gloom of one such chamber or space is traversed now by a golden shaft of light falling through some break in the upper foliage, giving a strange glory to everything it touches—projecting leaves, and beard-like tuft of moss, and snaky bush-rope. And in the most open part of that most open space, suspended on nothing to the eye, the shaft reveals a tangle of shining silver threads—the web of some large tree spider. These seemingly distant yet distinctly visible threads serve to remind me that the human artist is only able to get his horizontal distance by monotonous reduplications of pillar and arch, place at regular intervals, and that the least departure from this order would destroy the effect. But Nature produces her effects at random, and seems only to increase the beautiful illusion by that infinite variety of decoration in which she revels, binding tree to tree in a tangle of anaconda-like lianas, and dwindling down from these huge cables to airy webs and hair-like fibers that vibrate to the wind of the passing insect's wing.

Hudson's description is notable for its wonderful attempt to convey the incredible visual effect of a forest like that of La Selva. What it lacks is some sense of what it feels like to be there—the strange immensity; the feeling of being surrounded by so much life of so many different kinds; the hummings, rustlings, and susurrations; the strange cries, warbles, and whoops; the vibrant feeling of being in the presence of something huge and ancient and completely indifferent to your presence. And that's during the day. The sensual experience of walking through the forest at night is perhaps even more intense because the warm velvet dark envelops you while the croaks, trills, coughs, and stridulations impel a sense of being inside something immense.

Now try to translate these descriptions into the problem of migrant sampling that confronted us. In Veracruz rainforest, most migrants were captured at some time or other in nets set within eight feet of the ground. Even species like redstarts and Magnolia and Wilson's warblers found normally in mixed-species flocks were caught, even though the normal foraging height for these birds was fifty or sixty feet. Not so at La Selva. Migrants accompanying mixed-species flocks foraged eighty to one hundred feet above the ground and seldom were caught in our nets.

Clearly, our methods required immediate modification. We had to have a way to quantify migrant use of habitats, and mist nets by themselves were not enough. This sort of disaster is common enough in field projects. Methods that seem perfectly suited to address the questions at hand when considered from the comfortable setting of your office turn out to be quite useless when attempting to apply them in the field. Creative modification is often required but is painful at the time. Thousands of dollars and months of planning are riding on your decisions.

As a result of our sampling problems, we added an audiovisual censusing method to the mist-netting in order to make sure we were including migrants that occurred in the various habitats but were not being captured in mist nets. This method can be described as follows: At each sampling site, twelve points two hundred meters (650 feet) apart were censused. At each census point, all migrants detected visually or acoustically within a fifty-meter (160-foot) radius of the center point were recorded along with their estimated distance from this point.

This procedure allows you to calculate a density in a given habitat of birds heard or seen, serving as a valuable addition to sampling with ground-level nets because you can include those birds that spend their lives far above those nets. Nevertheless, its practical use in rainforest is challenging. The method was developed for use in open riparian forest bordering rivers in Colorado, where it is pretty easy to be sure that you have heard or seen every bird within seventy-five feet of your location. That assumption is simply not true in rainforest, where seventy-five feet includes a vertical element clogged with branches, bromeliads, and lianas. In addition, we are talking about *rain*forest. It *rains* there. A lot. In fact, it rained most days during our three months of study. This cuts down on your ability to see (through water-splattered and fogged binos) and hear (through the roar of a deluge on a billion leaves). But what alternatives did we have? None. So that's what we did. We added the audiovisual census to the mist-netting grid for every sampled site.

As I have said, Rafa and I began the work by driving all over northern Costa Rica trying to figure out what habitats were represented by what colors. Fine. Most of this kind of sampling could be done from the Suzuki. That took us a few days to accomplish. Then the hard work began. Once we knew what habitats we had and where they were found, we had to select three sites, 2.5 acres in size, in each major habitat type where we would set up mist nets for capturing birds, perform audiovisual censuses of birds seen and heard on the site, and conduct an intensive procedure for characterizing the physical structure of the vegetation—canopy height, ground cover, foliage density at different heights above the ground, and general description (such as mature forest, plantation, second growth, pasture with scattered trees, etc.). These sampling sites were supposed to be chosen using some sort of random selection process because if the sites are not selected at random, you cannot analyze the data collected from them using parametric statistics. To satisfy this requirement, we prepared numbered paper slips for each square-kilometer block from a topographic map representing the major habitat types, threw them into a pot, mixed them up, and chose three such slips for each habitat. Sort of. In truth, we had discussed exactly how our random sampling would be done with the USFWS statistician back up in Maryland in Mike Erwin's Patuxent office. There we had explained to this gentleman that true random sampling would be logistically impossible. Maybe a Delta Force commando can carry a sixty-pound pack across a couple of miles of trackless jungle in a half hour or so, but we knew that just getting to sites a mile or more from the road would be far too time-consuming for us, let alone performing all of the netting and other sampling required there over a period of days. So we explained that we would perform a "stratified" random sampling—which is to say, we would set some parameters to limit which sites would be included in the sampling lot. In our case, we limited the samples by requiring that they be located within a half mile of some road or access path. Our statistician told us that such a practice had no statistical validity. Random means random. As soon as you qualify the sampling in any way, it is no longer random, regardless of whatever euphemism we might wish to use in describing it. And he was right, of course. Limiting our random sampling to those square-kilometer blocks within a half mile of the road was problematic, but we had no choice. We had funding for three months total in-country, and all of our sampling in all of the habitats had to be completed within that time period. In field biology, you do what you can do, keeping in mind that our procedures might have skewed our results in some way.

The other issue for us was permissions. In addition to the vast cultural differences between Costa Rica and Mexico, the project that we were pursuing was different from any of those done in Mexico. In Mexico we did most of our work at a few sites, which greatly simplified permission requests and logistics. Once we had spoken to the land-owner and established our netting sites, we were set for the season. Not so for our Costa Rican project.

It is interesting to note that most countries in this world have no tradition what-soever of conducting scientific inquiry that is not directly related to some commercial or military enterprise. When foreigners visit for any purpose other than tourism, it is assumed that they are there for purposes of exploitation. This assumption is obvious at nearly all levels—from national, state, and local governments down to the individual landowner. Several times in various countries, I have been informed by some official or rancher or farmer that I was an obvious CIA plant. I have often tried to reason with them regarding what possible interest Central Intelligence could have in their cornfield or woodlot, but to no avail. It's a lot easier for them to believe in a mysterious plot than to accept that a team of outlanders could spend thousands of dollars and months of their time to learn about birds that aren't even theirs. In our years of work for long periods at specific places in Texas and Mexico, the locals eventually had to admit that however crazy it seemed, our hundreds of hours of work on natural history was our only apparent activity and that we were pretty harmless—maybe even giving a slight boost to the local economy. The situation was quite different, however, in Costa Rica. Here our procedures required us to pick places to do fieldwork at random on a more or less daily basis. Which meant that I had to do my best to find the landowner and try to convince him in my crude Spanish to give us permission to do the work—walking around on his property, cutting paths through his vegetation, setting up mist nets, and often chasing his cows to keep them from tearing the nets to bits. In a way our situation was similar to that of the migrants that we were studying, particularly the young ones. They too are confronted each fall with the task of finding suitable spots in a foreign land, "coming home to a place they've never been before," like the song says.

People were great to us in most cases. Even in those few instances when they were unwilling to let us abuse their vegetation, they would say something like "Come back tomorrow." But of course, we couldn't always locate an owner, especially in large areas of second growth or forest. Then we just went ahead, figuring it would be easier to ask for forgiveness should the owner appear. This worked for us for the most part, but there was one instance indicative of the problematic nature of conducting fieldwork in a strange land. It occurred during our third field season on February 18, 1989, at a swamp forest site located a couple of miles northeast of the town of Tigre. My field assistant at the time, Dave Swanson, and I were headed out from the netting plot in dusk at about 5:30 in the evening when we were accosted by two men who wanted to know what the heck we were doing there. Evidently, our comings and goings in Tigre had been the talk of that small community. People thought we were either "dangerous men" or engineers exploring for oil. We explained who we were and what we were doing, but I doubt they believed us, as we were just a mile or two from the disputed border with their war-torn neighbor Nicaragua. They seemed relieved, however, that we were *Norte Americanos*. As such, it was easier to believe us wandering idiots than hardened mercenaries.

A significant difficulty for us, in addition to that of the "random" selection of sites and obtaining permissions for their use, was the necessity of getting to them and setting them up. If access to the site and the site itself were pasture, cropland, or plantation, which was true for about a third of our sampling areas, then no problem. However, if site locations required passage through forest or second growth, that involved some hard work. Older forest with a closed canopy is not too bad. The canopy trees are spaced some feet or yards apart, and the undergrowth of saplings, vines, and fallen logs can usually be negotiated without great difficulty. Not so for younger forest and second growth. *Second growth* sounds pretty harmless. By definition, it is the vegetation that grows at a place where the "primary growth," primeval rainforest for the most part in our study area, has been cut. The problem is that the riot of plant life appearing when allowed access to the sun and water by tree removal creates within a year or two a tangle of limbs, leaves, vines, and spines from ground to canopy several feet in height. Confronted with this nasty wall, the word *impenetrable* jumps to the mind—but, of course, it is not. It cannot be, because your "randomly" selected site is located half a mile inside this mess, which means that you have to sharpen up your machete and cut your way through. About half of our thirty sites required significant passage through some form of this habitat.

I have mentioned that rain was a problem. But on November 21, it became a *big* problem. It was during that night that a *temporal* hit. This weather event has a cause similar to what we would call a *norte* in Veracruz—namely, a cold front dipping down across the Caribbean from the United States. However, unlike *nortes* in the Tuxtlas, there was no wind or temperature change. Just rain. A *lot* of rain. I quote from my notes below:

21 Nov 87——During the night it rained hard and continued to rain all day. George and I drove out to talk to folks about getting permission to net in their pasture. We looked at pasture near the La Selva station and a five-year-old second growth with help from Edwin Peniagua—caretaker of the La Selva second growth patch near La Gloria [a small village]. We examined several sites, and it rained and it rained. By 1500 [3:00 p.m.], we went to try and do a random habitat analysis. So we drove to Tosi's place to get the necessary equipment. The water was way up and the passage along the road a bit tough but still makeable. As we were sitting there considering, rain pouring down, we could see that the Sarapiquí was 20′ above its normal level and rising fast. It was within inches of running under Tosi's house! We got our gear together in a hurry and set off [to try to drive out]. What a sight. Our road in was now inundated completely, though we had been inside [Tosi's] less than an hour. We gunned it through [in the Suzuki], trying to make it, but one [flooded] stretch was 25 m long [80 feet] with water 1.2 m [4 feet] deep. It was too much for the Suzuki. We had no exhaust extension, and we didn't hood the fan belt. It stalled 10 m [30 feet] from safety. We (George, Rafa, and I) had no idea whether the thing was completely destroyed or not. Water filled the cab well over the seats, wetting quite a bit or our gear, though we were able to get some things up fast enough to keep them dry [GPS systems, maps, data sheets, etc. do not do well with water]. We got out through the windows and pushed the thing the rest of the way out [on to the "shore," more or less]. Then we debated whether to (1) drain gas, oil, etc.—refill with clean and start; (2) tow to mechanic; or (3) try it.

We then paused to consider our predicament. If the sinking of our vehicle had allowed water intake through the exhaust, as seemed likely from the immediate quiet of the engine, it was probable that the engine was dead, the cylinders distorted. But what if only a small amount had entered? Maybe if we drained out all oil and gas, let it sit for a few hours, and then replaced the fluids with clean stuff, it would be OK. Or maybe not. And of course, the only place to obtain said replacement fluids was from a gas station—the only gas station in the region, and it was fifteen miles away. And what

Figure 23-4. Suzukis float! Rafa and George considering our pickle from their respective perspectives.

if the engine was dead? Where were we to get a replacement? The nearest settlement of any size was Puerto Viejo, the same place as the gas station. Would there be a mechanic capable of replacing the engine? *And* where would he get the engine? It would have to be from San José. How long would that take? Or would we have to get the thing towed back to San José and simply drop it at the rental place, let the insurance cover it, and rent another? After a few minutes spent in this sort of circular, agonizing discussion, I decided that regardless of the Suzuki's status, it was pretty useless to wait. We might as well try the engine and see what would happen. I did, and it ran—whew! God bless Suzuki engineers!

Spending a few months traveling hundreds of miles on back roads (or "roads") is a good way to get a feel for the character of a country from my experience, and so it was in northern Costa Rica. It didn't take long for us to realize that Costa Rica was not remotely similar to Mexico: fewer people; fewer towns; better main roads better maintained; no careening trucks and buses; no sharing the highway with livestock; no soldiers, *policía, aduana,* or *federales*; little evidence of indigenous culture; and most importantly for us, no food. Of course, I don't mean literally "no food." There was food, but nothing

247

whatsoever like in Mexico. Keeping in mind that when I say *Mexico*, I mostly mean the states of Tamaulipas and Veracruz, my decade and a half of experience had instilled in me the belief that Mexico has the best and most widely available comfort food in existence. I am no gourmet, and I know nothing about "cuisines," but I have traveled a lot in this world and found nothing comparable. I suppose it has something to do with the wondrous richness of its history as well as its busy urban and rural economy, but for whatever reason, no respectable village of more than a few hundred souls will be without its restaurants open to the public from early morning until late evening, and no stretch of main highway without its all-night taquerias every ten or twenty miles or so. Fanny Calderón de la Barca, consort of the Spanish ambassador to Mexico, even commented on this wide availability of excellent food and drink for travelers from Veracruz to Mexico City in 1838, although it took her awhile to develop a taste for some of the fare. She found pulque, a drink made from the fermented sap of the maguey, disgusting, stating that "as nectar was the drink in Olympus, we may fairly conjecture that Pluto cultivated the maguey in his dominions." (She admitted having developed a taste for it some months later.)

Many countries can boast of widely available restaurants, of course, including our own. However, few (none?) can claim the uniformly high quality and variety of what is on offer throughout Mexico. My personal explanation for this is that an expectation of excellent meals is a critical piece of every true Mexican's heritage, regardless of which particular branch he or she happens to derive from. I found this out when working in the field with eight men, six of whom were Mexicans of various backgrounds (Mexico City, Mayan, Zacatecan). We were living in a trailer supplied by Mario's organization, INIREB, at a remote site on the Coxcoapan River and only went out once a week or so for provisions. I did the shopping for us on the first trip: cans of Dinty Moore beef stew, tuna fish, sardines, Spam, baked beans, a dozen loaves of Pan Bimbo (white bread, like Holsom), a couple of jars of peanut butter, coffee, and a few packs of "Marias" (cookies). Nobody said anything, but my Mexican crew insisted politely that they would do the shopping in the future, and also the cooking. Which is when I found out that although Mexico has a strong patriarchal tradition, the men all know how to cook, and they know how to do it well. Given a choice between canned beans and sardines for breakfast and dinner for a month in the field, they will happily head to the nearest market whenever necessary to get the rice, beans, tortillas, onions, tomatoes, chiles, and so on required to concoct a decent meal.

Costa Rica does not have that tradition—at least, not outside the big cities. There were only two restaurants open to the public that we found in our entire region: one in the town of Puerto Viejo and one in La Virgen. Both were open for dinner only, and both had one item on the menu, a dish called *gallo pinto*, which is beans and rice.

Well, so we spent three field seasons in northern Costa Rica—November 1987, November 1988, and February 1989—studying migrant habitat use and attempting to quantify amounts of the habitats. What did we find? In terms of the results for answering our research questions, it all worked pretty well, as reported in our chapter in the Smithsonian symposium on the topic (Hagan and Johnson 1992). While netting and audiovisual sampling underestimate migrant use of rainforest, these methods provide fairly accurate assessments for tree farms and pastures. Also, the satellite, despite its obvious mistakes, could show that there wasn't much rainforest or even second growth and that most of the thirty-nine species of migrants recorded greatly preferred those habitats to the pastures and croplands that occurred at orders of magnitude larger amounts.

The toll taken on us and our field crews (Harriet, Rafa, and Dave Swanson during our third season), however, in terms of doing the site establishment, netting, and vegetation analyses in rainy, hot, and bug-infested sites (mosquitos in the forest, chiggers everyplace else) was significant. And the audiovisual sampling attempted by George and me proved to be challenging, to say the least. Visiting equatorial rainforest is a spiritual experience, but trying to pick out the migrants in mixed-species flocks foraging a hundred feet above you in a deluge all day, every day is no fun. It wears on you. In the end, I was glad when we finally completed our third and final field season there in February 1989.

One reminder of working in tropical environs came some weeks after my return when I gave birth, via modified Cesarean section, to a screwworm. Our family doctor, Carolyn Franklin, exclaimed "It's alive!" with surprise and horror on removal of the beast from my chest.

WORKING THE WAR ZONE IN BELIZE

Chandler S. Robbins, known as "Chan" to his many friends, fans, and colleagues, served as a USFWS biologist at their offices in Patuxent, Maryland, for over sixty years, retiring from that position in 2005. He made many contributions to the field of ornithology, among the most important of which was the origination and development of the annual North American Breeding Bird Survey (BBS), a countrywide count of breeding birds conducted each spring by a vast army of volunteers. The first count was in 1966, and it continues up to the present. USFWS organizes the volunteers, collects the data, and performs the annual statistical analyses that determine the population trends for the members of our avifauna.

By the mid-1970s, the BBS data showed a disturbing downward trend for many species. Initially, Chan and many other scientists attributed these declines to breeding habitat loss or fragmentation (breaking larger blocks into smaller bits). However, by the mid-1980s, it was evident that these explanations were not sufficient. The sharpest losses occurred in long-distance migrants like the Wood Thrush, which breeds in eastern forests in the United States and winters in the tropical lowland forests of Mexico and Central America. Yet breeding habitat for these birds in reality was *increasing* and had been since the early 1900s, whereas scientists working in the Neotropics reported the literal decimation of lowland forest throughout Middle America. These data suggested that migrant population declines might be a result of wintering habitat loss rather than that of breeding habitat.

The issue was not academic for Chan. He had observed firsthand the destruction of beloved forested areas in Maryland and had no doubt whatsoever of its devastating effects on birdlife. Nevertheless, he could understand that the migrant life cycle might make them vulnerable to more than one issue. With these ideas in mind, he got some money from his parent organization to begin an investigation of migrants wintering in

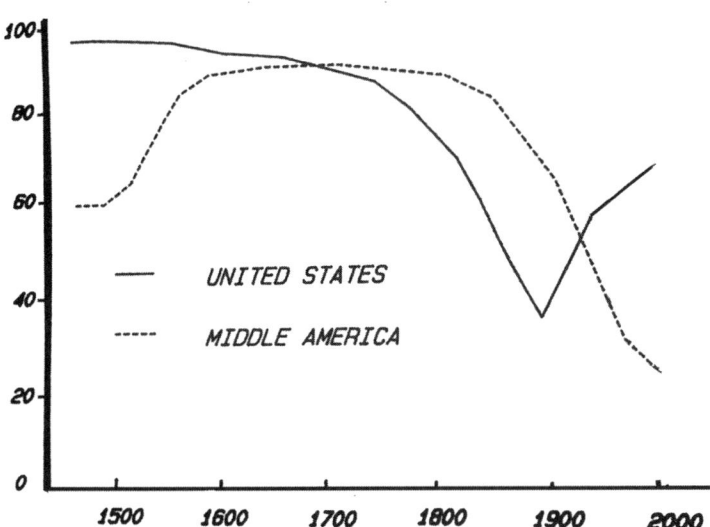

Figure 24-1. Forest cover in the United States and Middle America over the past half millennium. Data from the United Nations Food and Agricultural Organization.

Belize, where he had some good in-country contacts that would simplify logistics. Chan and his team began at about the same time that George Powell and I were doing our work on migrants in Costa Rica. We were, in fact, getting our funding from Chan's USFWS parent group at Patuxent. Thus it was a natural extension of our Costa Rican research to move into Belize, where our use of satellites for habitat assessment could provide valuable data regarding the actual amount of the habitats that Chan and his crew were sampling. So when we finished up in Costa Rica in early 1989, we began discussions with Chan about doing similar studies in Belize. By the time fall rolled around, we had obtained from our rocket scientist, Steve Sader, an unsupervised classification map of the southern portion of Belize, where most of the lowland tropical forest remained, and were finalizing plans to get the project underway.

The near-death experience of the Suzuki in the Sarapiquí convinced me that I should bring the 1979 Chevy Blazer (the famed pig transporter) purchased for one of my Mexican projects down to Belize from South Texas for the planned three-year duration of our project. On November 4, 1989, at 3:15 in the afternoon, Rafa and I cleared customs in Matamoros and set out on the thirty-six-hour trip through the Mexican states of Tamaulipas, Veracruz, Tabasco, Campeche, and Quintana Roo to the border. I have

Figure 24-2. Central America with Belize highlighted.

described elsewhere (chapter 10) our various issues encountered along the way during the eleven stops by local police, state police, drug police, customs, *federales*, and the army. It was a tough trip. Nevertheless, we arrived in Chetumal late in the graveyard watch and got a room at the Paradise Hotel for five hours of sleep before crossing into Belize at eight the following morning.

Belize is different—different from Mexico, different from Costa Rica, different from anyplace else. Really different. It was part of the Mayan Empire until the collapse of that entity roughly a thousand years ago. Mayans of three groups remain the principal indigenous occupants of the country: Maya (or Yucatec), Mopan, and Q'eqchi'. Together, they constituted about 10 percent of the population when we were there in the early 1990s, living mostly in the interior hinterlands as subsistence farmers. People of African descent constituted the largest and most politically powerful faction in the country at that time. When I dealt with officials regarding my permits and other government issues, most were of this ethnicity. This situation, however, appears to have altered. At present, Mestizo people of mixed indigenous European ancestry constitute the largest ethnic grouping, at more than 50 percent, presumably due to immigration from neighboring Guatemala, which has doubled the population from about two hundred thousand in 1991 to over four hundred thousand at present. I

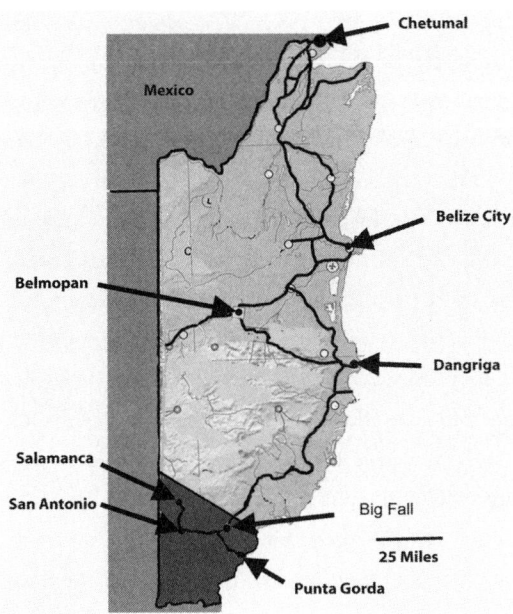

Figure 24-3. Map of Belize showing principal cities and our study area (shown in dark gray) in the south.

presume that this sharp shift has resulted in a comparable change in the ethnopolitical structure.

The Mayan occupants of the country at the time of colonization have had little to do with its subsequent history. The Spanish Empire laid official claim to all of Central America dating from the 1494 Treaty of Tordesillas, which ceded all newly discovered lands west of a north-south line drawn 370 leagues (about 1,300 miles) west of the Cape Verde Islands; the new areas east of that line, basically the Brazilian bulge and all of Africa, were to belong to Portugal. No countries other than the two signatories ever paid much attention to this line of demarcation, yet the treaty retains traditional if not legal importance.

After the Spanish conquest of Mexico and Central America in the early 1500s, the region of modern Belize was included as the easternmost part of the Department of Guatemala. However, the Spanish never had a strong colonial presence in Belize, which was dominated early on by people of British heritage who began settling along the coast in the mid-1600s. Spain attempted to enforce their understanding of ownership in 1798, when a flotilla was sent against the colonists. The defeat of this force at the battle of St. George Caye by the colonists (who referred to themselves as "baymen") ended Spanish claims. Great Britain recognized the colony as an official part of the British Empire in 1862, calling it British Honduras. It became an independent country in 1981. The origin of its name, Belize, remains a subject of debate, with some claims being more colorful than others.

Guatemala has never recognized the existence of the country as either British Honduras or Belize. They stake their claim on the original establishment of the Federal Republic of Central America as an entity independent of Spain in 1821, which included the region now known as Belize. As mentioned in the account regarding Costa Rica, the Federal Republic of Central America never functioned as a nation. Guatemala established itself as an independent country in 1847, including all of Belize as part of its national territory. The Guatemalans concluded a treaty with the British Empire in 1859 that recognized the territory of Belize, then known as British Honduras, as a British protectorate in exchange for British construction of a road from the coast to the Guatemalan interior. The road was never built, and this breach of a key provision of the agreement has served as the main claim to Guatemalan sovereignty over Belize since. The British maintained a military presence in Belize even after the country's independence in response to on-again, off-again threats of Guatemalan invasion. Official maps of Guatemala do not recognize Belize's existence. Referenda in both countries taken in 2020 support adjudication by the United Nations International Court of Justice. A decision, delayed by the pandemic, is awaited.

At 10:30 in the morning on November 6, 1989, Rafa and I drove the Blazer across the Río Hondo, and the great Belizean adventure began. Customs, thankfully, was no problem, and we were soon on our way. What I knew about Belize at this point came from two main sources: (1) a member of Chan Robbin's team who had done fieldwork in the exact same area of southern Belize where we planned to do our work and (2) the highly personal account of my dear friend Jonnie Fisk in her book *Parrots' Wood*, which recounted an episode of her time with a group of bird watchers doing work banding birds at Dora Weir's place on the road from Belize City to Belmopan in 1985.

I had taken the opportunity after one of our Patuxent meetings to question Chan's person in detail with specific inquiries concerning what I would have to prepare for as we set ourselves up for field seasons over the next three years. The questions and their answers are pertinent, I think, if only to remind one of the need for more than one source on any potential enterprise of importance:

Q—Is it possible to obtain topographic maps from any government facility in the country?
A—No.

Q—How are the roads into the southern part of the country?
A—Impassable for two-wheel drive, difficult for four-wheel drive.

Q—What language is spoken by the rural people whom we are likely to meet in the southern part of the country? Do they speak Spanish?
A—They do not speak Spanish. They speak a language of the Mayan family of languages called Q'eqchi'.

Q—What is the attitude of rural people to the presence of foreigners?
A—Unfriendly and mistrustful.

Q—Are telephones available for calls in country and internationally?
A—No. None outside Belize City.

Q—Is gasoline widely available?
A—No. None after Dangriga.

Q—What about food?
A—You will need to stock up in Dangriga.

Q—Are there inns available in the southern regions?
A—No. You will have to camp.

Q—Are there ways to travel to the southern regions other than by vehicle?
A—No.

Based on this detailed information, I made my plans accordingly, purchasing a Q'eqchi' dictionary and whatever poor tourist maps I could find, driving down my own four-wheel drive from Texas loaded with gas cans and camping gear, stocking up with supplies in Dangriga, and so on.

Fortunately, as I will relate item by item, nearly all of this information was completely wrong, leading me to wonder what country this person had visited. We began learning this truth within an hour or so of clearing customs when we stopped for gas at Corozal. Questioning the attendant, I found that Spanish was indeed not common except among Guatemalan refugees. However, English was the principal tongue, spoken by most people regardless of ethnic background. He said that although the roads were not great, they were all-weather and passable all the way to Punta Gorda at the southern tip of the country, which, he said, was about a fifteen-hour drive from where we were standing.

Heading out from there at about noon, we got to Orange Walk a couple of hours later, from which point I called George (telephones were readily available in any population center, even for international calls). He had been scheduled to fly into the Belize international airport outside Belize City the next day but had decided that he had too much work to do in Florida and would not be able to get free to come down to help us with the fieldwork. Disappointing, but not entirely unexpected.

We arrived in Belize City at about 3:30 in the afternoon. I was able to find the post office after a bit of exploring, where I was in hopes of obtaining topographic maps of the country. They had them all right, but none of the Toledo District that I needed. They advised me to check at the land office in Belmopan. I had thought of heading the fifty miles there directly, but the front wheel bearings had begun making ominous clicking noises, so we decided to stay in town for the night.

Belize City in 1989 was not a tourist town. Most visitors from other countries came for the "cays," islands in the Caribbean just off the Belizean coast justly famous for their gorgeous beaches and crystalline seas. These people would fly in from New York or LA into the airport north of the city near Ladytown and then take an air taxi out to their luxury hotels on the islands, completely bypassing the colorful experiences of the city.

Ramshackle seemed to best describe the place for me—a haphazard collection of mostly wooden structures in need of repair bordering a series of canals that drained the city of its waste. A third of the people in the country lived in this population center when we were there, and based on my own experiences, rather a larger portion of them than seemed right were intensely interested in providing me with any services I might want or even imagine.

When we went to supper at New Archies, a Chinese restaurant near our shabby hotel, we were barely seated before a casually dressed young man sat down, uninvited, to join us at our table. He used a long list of questions to get to know us better, and then, feeling more relaxed, proceeded to advise us regarding potential delights available for persons of modest means, mostly chemical from him, but for a small fee, he could put us in touch with colleagues who could help satisfy other needs. Eventually recovering from my surprise, I explained that we wished to dine alone, but I had to start getting out of my chair before he took the hint that we were grumpy, not the fun seekers he had hoped, and departed.

The next day, November 7, 1989, promised to be a busy one: truck repairs, permits, maps, and money exchange were all yet to be accomplished. It didn't start well. Hot water is a luxury in the inns of many countries, but water of some sort is generally expected as part of the service. Through some unexplained problem, our hotel lacked this basic commodity. So we set off unwashed on our busy schedule. On our way to the truck, we noticed a couple of men trundling what seemed to be a white sack down a stairwell before dumping their load unceremoniously onto the sidewalk. The "sack" turned out to be the body of an old woman, as sobering a memento mori as one could wish.

Things went rather better than I had hoped at the mechanics', where they not only were able to get right to work on our problem but had the parts necessary to complete the repairs, a rare event in my long history of such dealings. By noon, we were all paid up and ready to go, the only problem being that the "click" made by defective bearings had been replaced by a screech. The mechanics had no further advice to offer other than earplugs, so we set off for the bank to change money (two dollars Belize for one dollar US) and then headed to Belmopan, arriving there at about 2:30 in the afternoon.

Belmopan is as different from Belize City as can be imagined. The old capital was devastated by Hurricane Hattie in 1961, and plans were made to build a new capital inland, which were initiated in 1970. The avenues and main government buildings were completed and occupied by the time we got there, but only about four thousand people actually lived in a place built to accommodate several times that many. The effect was something like the *Twilight Zone* episode where the protagonists stroll through a town completely devoid of life, even flies—or maybe like walking down the broad, deserted, temple-lined thoroughfares of Teotihuacán. Although the government offices had been moved officially to the new capital, most ministers and bureaucrats still lived in the old capital. So while the avenues were lifeless, the offices hummed along within. We were able to purchase the majority of the topo maps that we needed in short order at the land office and went on to the Ministry of Agriculture, Forestry Division. Here things went a bit less smoothly but still perhaps better than expected. The problem for the Belmopan bureaucrats was that things were still so new that the normal barriers between those desiring a government service and those providing it were not yet completely in place.

The Smithsonian Office of Personnel is an excellent example of how the system is supposed to work when well established. When I was hired by the Smithsonian, my first paycheck was held up because of a lack of paperwork, which was eventually delivered to the Personnel Office a week or two after I had arrived and begun my new job. However, a month later, I had still not received a paycheck. I had called their office repeatedly, but all calls were answered by a machine requesting me to leave a message and my number and someone would get back to me. No one did. I had also tried going directly to their offices in L'Enfant Plaza in DC. Forget that. There is a reception room and reception-ist, but the only unlocked door out of the room is the one through which the visitor entered, and the receptionist will not allow anyone into the inner sanctum who does not have an appointment. The bureaucrats are perfectly insulated. Or almost. Out at the Smithsonian Conservation and Research Center in Front Royal, Virginia, where I was research coordinator, we had our own representative of the Personnel Department, Mary McComas. Mary had worked for the Smithsonian for a long time and knew its ways, so when I went to her with my problem, she said, "OK. Let's go visit them." We did, driving into DC in the third week in December, a notoriously bad time of the year to try to get anything done. We got there and went up to the receptionist's office. No one was there, so Mary simply banged on the door to the office section until someone opened it. Her tall, imposing form was recognized immediately, of course. She was there practically every week for something or other. Not surprisingly, the Office of Personnel was in the midst of a great Christmas party, which we were invited to join. Mary asked, "Where's Sally Brothers?" (All names changed to protect the innocent.)

"She's over there by the punch bowl," was the reply.

So Mary walked over to the woman and said, "Sally, this is Dr. Rappole. Where's his paycheck?"

Sally, taken aback, stammered, "I . . . I . . . I'm not sure."

So Mary said, "Well, let's go take a look."

Whereupon she went into Sally's office, over to her desk, and began shuffling though papers there. It took her about two seconds to find my paycheck. She stood up and said, "Thanks, Sally. Merry Christmas," and we left.

Such protections were not in place in the Belmopan government offices in 1989, so I just walked into the majestic structure of the Ministry of Agriculture, checked the sig-nage in the lobby for the Division of Forestry, and went on up to their offices. I asked for the chief of forestry, Mr. Rosado, but was informed that he was not in. Well, OK, so what about his number two, Dr. Belisle? No. He wasn't in either. What exactly did I want? A permit to do scientific research in the country, I explained. They knew nothing about it despite my correspondence, so they suggested that I go to the Ministry of Labor and get a work permit. So off I went to the Ministry of Labor, where I was told that a work permit was ridiculous. No one was paying us to do anything, so why would we need a work permit? Fair question. So I went back to Forestry. Mr. Rosado was in but unhappy about the immediacy of my request for a scientific permit. For instance, where was the proposal? No worries. I walked out to the truck and wrote up a brief description of our proposed activities and their purpose and went back in at about 4:00 p.m. This time, Dr. Belisle was in, who was, in fact, in charge of permits.

Dr. Belisle was surprised that Mr. Rosado had made us wait but said that in addition to the proposal, we would need letters of support from the sponsoring agencies, the Smithsonian for me and USFWS for George. Once those were in hand, he felt the issuance of a permit would not be a problem. Whereupon we left and got rooms for the night at the Bullfrog Inn. I called Bon from there and asked that she contact my boss, Chris Wemmer, at the Smithsonian Conservation and Research Center and have him send a letter, the fastest way possible, to Dr. Belisle and to do the same for Mike Erwin at USFWS.

The next morning, we set out for the south early—two hours to Dangriga, then two hours to Maya Creek, then two hours to Big Falls, then two hours to Punta Gorda. Along the way, we learned of some additional errors in the information conveyed to us back at Patuxent:

> Gas not available south of Dangriga—Not so. There was a gas station at Big Falls in the heart of our work district.
>
> Supplies not available south of Dangriga—Not so. Supplies were readily available in Punta Gorda, the principal town.
>
> Accommodations not available south of Dangriga—Not so. There was a small hotel in Punta Gorda, the St. Charles Inn, owned and run by two brothers, Heston and Dwayne Wagner. There was also what you might call a "tourist court" available in the Q'eqchi' Mayan village of San Antonio, although I did not find that out until later in our stay.

It was true that Mayans in the rural Toledo District did not speak Spanish. Fortunately, however, they did speak English thanks to pretty much universally available public schooling sufficient to learn reading and writing in that language.

Finally, we found no evidence of the unfriendliness of the local people in the country of which we had been warned. Once outside Belize City, our experiences were entirely positive and, for the most part, helpful, even if our presence and mission seemed completely incomprehensible. The only stiffness we encountered was from those managing properties for large landowners. These people, while polite when considering our requests to net and observe birds on their employer's property, were understandably reluctant.

By November 8 we were set up in lodgings at the St. Charles Inn in Punta Gorda. They had a telephone available there, and I kept it pretty busy for a while, making calls (1) to Bonnie to make sure things were OK at home and find out if she had gotten through to Chris and Mike about supporting letters, (2) to Dr. Belisle to let him know where we were and that the supporting letters should arrive soon, and (3) to Mr. Edwards, chief forest ranger for the district, to let him know what we were up to. I had found out from our innkeeper, Dwayne Wagner, that there was daily taxi plane service from the municipal airport in Belize City to Punta Gorda for $90. I thought this information might convince my partner in this endeavor to fly down and join us, so I called George as well.

The next day, November 9, we set out to work on verifying our unsupervised classification from Steve Sader's Landsat satellite map of southern Belize, which showed the following map colors and habitats:

> Dark green, green, light green, dark brown—Older second growth and young forest (more than thirty years in age, based on core samples)
> Yellow, purple—Older second growth with some trees (five to thirty years old)
> Red—Young second growth (less than five years) and overgrown pasture
> Pink, gray—Fallow fields
> Blue—Pasture
> White—Bare ground
> Light brown—Pine plantation

Unfortunately, the one fact that our Patuxent informant was correct about was the lack of primeval, or even mature, forest. Most forest patches we were able to visit had evidence of cutting within the past fifty years or so. Eventually, we did get deeper into the Maya Mountains and found forest that was ancient, if not primeval. There are paths into these areas, but no roads. Still, I doubt there is much "untouched" lowland rainforest there. For starters, only the river valleys are lowlands. Most of the land has elevations above a thousand feet, which is outside the elevational range of lowland rainforest. Also, this part of Central America has been occupied by civilizations for thousands of years. These societies likely took advantage of forested lands in relative proximity to their communities for timber. Certainly, the English who settled along the coast in the 1700s and 1800s likely did, using their slave-based economy to harvest and export wood to timber-hungry countries in Europe. The treaty drafted with Guatemala in 1859 deeding the lands now known as Belize to Great Britain was agreed to based on Great Britain's construction of a road through the Maya Mountains connecting inland Guatemala with the coast, presumably because of the high value of the timber market.

In any event and for whatever reason, we found little primeval forest anywhere near what we could drive to, although we were assured that *high bush*, the Belizean term for "mature forest," still existed within a day or two's walk from San Antonio. This information we found to be correct when we had the time to walk back into these more remote areas.

Most of Belize is in what Leslie Holdridge, the dean of world plant community identification, would define as a "tropical moist forest" life zone. Only the extreme southwestern portion of the country harbored, or had harbored, rainforest. Our search for primeval rainforest in the southernmost fringes of the Maya Mountains in the Toledo District of Belize was complicated. Green was the color associated with forest for our satellite maps, but this term covers quite a bit of ground. It is likely that lowland forest in this part of Belize—that is, below a thousand feet in elevation—was indeed what Holdridge would classify as "tropical rainforest" prior to human modification. However,

Figure 24-4. The Belizean flag.

humans have been modifying this habitat in southern Belize for a long time. Mayan culture reached its peak here about eleven hundred years ago, and there are plenty of as yet unexplored and undocumented Mayan temples throughout the region. There were three such temples in the vicinity of one of our second-growth sites, and we found one on our Salamanca rainforest site, all unexcavated. Certainly, the Mayans didn't keep the trees around when they were building their civilization, so one must assume that all but the steepest slopes were under some form of cultivation during that period. The Mayan civilization had collapsed by the time of the Spanish conquest and colonization, and so rainforests likely covered what is now southern Belize once again by the time of British colonization in the mid-1600s. A principal source of income for these colonists was timber exports, chiefly mahogany; in fact, it is worth noting that the Belize flag shows a mahogany tree fronted by two loggers, one Black and one Mestizo, with the motto "Sub umbra floreo" (Under the shade, I flourish).

I don't know how this market affected forests west of Punta Gorda, but it seems likely that timber was harvested from this area, as from everywhere else in Belize, at least up until 1984, by which time most trees of value had been taken and a moratorium on exports was put in place. It is unlikely that there is any "primeval rainforest" left in the country, if you mean untouched by humans—and indeed, most of the places that we went to had obviously been touched by humans a lot. Nevertheless, we did find some where there were no obvious signs of cutting. But how long does it take for obvious signs to disappear? I'm sure that a tropical plant ecologist could tell me, perhaps based on height and species composition of the dominant trees. And maybe it doesn't matter from the perspective of the bird community.

On November 10 (my forty-third birthday), we began our investigation of the various habitats available in the region covered by our map. Initial surveys of the young, disturbed forests that we could find revealed many of the same migrants found elsewhere in Middle America in forest and second growth: Wood Thrush, Kentucky Warbler, White-eyed Vireo, Yellow-bellied Flycatcher, Black-and-white Warbler, Philadelphia Vireo, Summer Tanager, Magnolia Warbler, Wilson's Warbler, Louisiana Waterthrush, Gray Catbird, Yellow-throated Warbler, Common Yellowthroat (marshes), and Yellow Warbler (pasture with scattered trees).

A striking aspect of these first surveys was the almost complete lack of any monkeys, squirrels, agoutis, coatimundis, pacas, anteaters, or large or even medium-sized birds like toucans and oropendolas. The likely reason for this paucity became clear during one of my first discussions with a landowner concerning access to his property, basically overgrown fields with scattered hedgerows of five- to ten-year-old second growth. He said, sure, we were welcome to poke around, but we should take care. He had shot a large male jaguar that was eating his chickens just three months ago. This story is not shocking. Any rancher in Virginia will explain how he has shot at least one mountain lion and a couple of wolves stalking his stock within the past year or two, despite the lack of any confirmed records for the state in the last century and a half. So it was not the story that was surprising but the denouement—the shooting of the predator. In Mexico, guns are illegal, except for "sportsmen." As a result, they are relatively scarce, maybe one or two beat-up shotguns per small village. Also, ammunition is not readily available, and so it is not wasted on species that don't do serious damage or provide a reasonable meal, so decent populations of medium-sized diurnal mammals and birds can still be found. Not in Belize. As I was soon to discover, nearly every subsistence farmer had an old firearm of some sort and apparently ready access to bullets and shells—hence the lack of game. Now I'm sure there are laws protecting wildlife in Belize, but laws require enforcers, and I can verify that my teams and I encountered few government officials of any kind during our months of work in the country, let alone game wardens. Larger wildlife species likely persist in woodlands deep in the Maya Mountains, where hunters have less ready access. After all, Alan Rabinowitz was able to capture and radio-collar six jaguars in the country for his work with the Bronx Zoo's Wildlife Conservation Society in the late 1970s. Nevertheless, and for whatever reason, most of the country away from the highlands is depauperate in edible mammals and birds.

I had another encounter later that day that has since stuck with me—but for a different reason. Driving south from the village of San Antonio, I found a good-sized block of fifteen- to twenty-year-old second growth where I decided it would be good to have a netting plot, so we drove on to see if we could find someone who could grant us permission. A few hundred yards farther down the road, we found a tiny clapboard shack with a corrugated tin roof. A man was seated on a bench by the open door. He appeared to be Mayan, perhaps fifty years old, dressed in shorts and an open short-sleeved shirt and thong sandals. I stopped the truck, got out, and approached. I introduced myself, and he responded courteously, saying he was Nelson Mandero. He then offered me a seat on a nearby bench and a glass of water, which I accepted. He listened to my description of proposed activities and request for access (in English, of course).

When I had finished, he began his own story. As to the land, he explained, he no longer owned it. He had sold it along with the fine house, fertile fields, and fruit and vegetable gardens that he and his wife and children, when they were younger, had cared for and nourished so diligently over the years. Now all he had was this shack, a former outbuilding. Everything else was gone. Even the hut was not his own.

As to how this dissolution had all come to pass, he explained simply, "After my wife died four years ago, I got discouraged."

This final word was freighted with such pathos that it hit me like a blow, an almost physical impact. Well, there was nothing more to say. I got up, thanked him, and left, like Coleridge's wedding guest—wiser perhaps, but certainly deeply sadder. That scene and those words remain clear in my mind more than thirty years after the event.

Sometime later, while Rafa and I were working on censusing the birds on Mr. Mandero's former holdings, an affable young man—in his early twenties, I would guess—came up to us. He explained that he had learned from his father what we were up to and that the land belonged to him. The piece was about forty acres in size, and we were quite welcome to use it. In fact, he would be more than happy to sell it to us. I responded with some surprise that I thought it was illegal in Belize to sell land to foreigners. He said that any such concerns could be easily dealt with without problems, and the land would be ours—at a reasonable price too, a few hundred dollars, if memory serves.

The next day, November 11, 1989, was a fateful one for me. It started out normally enough. I phoned Bonnie from the St. Charles Inn at about 8:00 that morning and asked her to call George and let him know that if there was any way he could get free for a few days, it would be a great help to us. He could fly from Miami to the Belize International Airport outside Ladyville, take a taxi from there to the municipal airport in Belize City, and an air taxi from there to Punta Gorda. Flights were relatively cheap, and he could be in the bush in a matter of hours. I didn't think he would come, but it was worth a try.

With that done, we went to eat at the café at the little airport just outside Punta Gorda—coffee, omelet, and toast. It was, in fact, the only place where a member of the general public could find breakfast in Toledo District back in those days, so far as we were able to discover. Like Costa Rica, and many other countries, there were simply not enough visitors or middle-class clientele with disposable income outside of large population centers to make a breakfast place worthwhile.

We then shopped for a few items at the hardware store in town (hammer, nails, twine) and headed out to set nets in what looked to be decent mature rainforest in the rugged mountains not far off the track between Blue Creek and San Antonio. We got to our access point at about noon, which was through a well-tended farm with a handsome house, barns, livestock, and croplands belonging to Alf Koon and his family—as tidy a place as one might find in Iowa or my own home in western New York. They were Mennonites, members of a religious group doctrinally related to the Amish. Many members of this sect had emigrated, mostly from Germany and the United States, to various places in Central America for the cheap land, low taxes, and freedom to practice their religion and raise their children as they pleased. There were about six thousand Mennonites in Belize when we were there, a number that has since nearly doubled.

Alf was a pleasant guy. He offered us some orange juice and was happy to chat for a bit. He readily granted access to the forest but said he didn't own it. He didn't know who owned it, maybe the government, but he was sure that it was not virgin. In fact, he was certain that there was no primeval forest left in Toledo District outside the interior of the Maya Mountains, and perhaps not there either. Everything had been at least high-graded, with the removal of all timber of significant size with commercial value within the last half century or so, and probably more recently than that. Nevertheless, he doubted anyone would mind if we put up our nets in whatever type of forest remained in the unlikely event that they even knew we were there.

With that encouragement, we packed up and headed for the bush. To get to the forest, we had to cross a mile or so of ten-year-old second growth. This task was no joke, and I wished, as I had many times previously, that our USFWS statistician from Patuxent could have been with us to share the experience of accessing a "randomly located" sampling point. This particular "second growth" had a canopy height of about fifteen feet but was only seemingly impassable for the first five or six feet above the ground—a wall of brambles, vines, branches, and other miscellaneous vegetation so dense that light could hardly penetrate to the forest floor.

This seems like a decent segue into a brief discussion of the remarkable tool—or weapon, as the case may be—that allowed us passage. Machetes come in many different types, each best adapted for use in cutting specific kinds of vegetation: long, broad machetes for grasses; short, thick, heavy machetes for dense branch tangles and bamboo; and so forth. Our machetes were of the type known as a "bush machete" here in the States. It is a general, all-purpose machine for cutting through mixed young brush. The blade is about two feet in length from grip to tip and two inches in width, with the cutting edge thinner than the support edge and rounded up to the support edge at the tip. There are two additional items essential to the user that no good brush cutter would be without: an eight-inch triangular file and a stick about sixteen inches in length and an inch thick with a four-inch stub at the end angled back toward the shaft at about forty-five degrees. When cutting through dense brush, the protagonist wields the machete in the right hand and the stick in the left hand, using the stick to pull branches and tangles down and to the left, exposing and placing tension on the stems and allowing an angled slice with the machete hand. Using a properly sharpened machete correctly wielded in tandem with the machete stick, an adept can whack his way through apparently impenetrable tangles at surprising speed. Not having any such persons with us, it took us four hours of hacking to make our way through the mess, the last bit sharply uphill as we climbed the lower flanks of the mountain.

Once we arrived, we found the forest at our destination to be disappointing. Our satellite map showed green, and there were trees, some as tall as seventy or eighty feet, but they were obviously young, probably none more than fifty years in age. Also, they did not make a closed canopy as one would expect in mature forest, allowing more light to the forest floor, and therefore more undergrowth, than was really right. But it was what it was. This disturbed woodland was what passed for "lowland rainforest" in the area covered by our map that was anywhere near a road, and we would just have to find out how well migrants liked it in comparison with the younger second growth, plantations, pasture, and cropland available.

We set out to reconnoiter the site for a possible netting grid, and it was during this effort that I made a serious error. I had left my machete stick back with my pack where we had first entered the forest and was using my left hand to pull away and expose vines in my path for cutting. This is a tyro mistake. The machete stick not only exposes and tensions the branch or vine; it also pulls it away from the body so that the follow-through of the slice endangers no body parts. As I was making my way through a tangle, I grasped a vine in my front perhaps an inch in diameter and swung to cut it out of my way. The force of the blow brought the vine down toward me, and the blade sliced through the vine and my Levis and into the top of my left shin just below the kneecap.

The wound seemed to bleed more than was right to me for bone and tissue, but it was time to leave anyway, so I wrapped it as tightly as I could with a bandana, and we headed back to Alf's, where we had parked the truck. The trip went a lot quicker than on the way in, now that we had a path. Still, I was pretty tired and light-headed by the time we got there. Alf met us at the truck and argued with concern that I should probably let him bandage things up as best he could and spend the night with his family at the farm. I thanked him and said I thought we had better take the hour drive back to Punta Gorda.

So we set off with Rafa driving and arrived in town in the early evening. We knew nothing whatsoever about medical facilities, so we went to the Chinese restaurant, the Kowloon, the proprietress of which was a friend of ours who knew everything and everybody of importance in town. She said we should head to her friend Dr. Moreno, which we did. He took me in and started to work, but when he endeavored to remove the blood-encrusted kerchief from the wound, he must have ripped open the small artery that had been just partially severed. In any event, the blood began to spurt with my heartbeat. He jumped up and said we should head to the hospital emergency room immediately. So Rafa and I got in the truck and followed his directions to that establishment. I was feeling a little shaky by the time we got there, but the emergency room doctor, an East Indian, greeted us calmly, inspected the wound, and said that if I would just remove my boots and clothes and get up on the gurney, he would get me fixed up in short order. I did as I was told, but after clambering onto the gurney, the next thing I was aware of was the doctor on top of me giving chest pumps.

After that, things calmed down. He got the artery clamped off, stitched me up, and started an intravenous line in my arm with a saline solution—no blood, thank goodness (this was at the height of AIDS, and contaminated blood supplies were everywhere). Then they wheeled me into the men's ward for the night.

This was a large rectangular room, maybe twenty feet by forty feet, with ten beds or so spaced around the perimeter. The nurse in charge—a tall, businesslike Black woman—did her best to make me comfortable, but it was not going to be a restful night. At least three of my neighbors looked and sounded like this might be their last moments on earth, and at about one in the morning, a drunken Mayan was wheeled in, bandaged for a knife wound in the chest. Despite the obvious seriousness of his wound, he had a lot of energy and a lot to say, most of it apparently in Q'eqchi' at a high volume. What happened to him during the night, I was not quite sure, but his bed was empty when I awoke the next morning.

The doctor came by early, checked the wound, and said that he thought I could leave if I felt up to it. I said I would be delighted to escape and asked what I owed. He said that it was a government facility, so there was no charge. I said that since I paid no taxes, I would donate what I thought his service was worth to me and gave him $150 along with my heartfelt thanks. When I went to dress, I found that my bloody boots, a nice pair of Timberland Pros, had disappeared. Oh well. Things could have been a lot worse.

Rafa drove me over to the hotel, and I hobbled up to my room. I hadn't gotten much sleep the previous night and so decided to rest there until supper time. Videos of the wall coming down in Berlin were on the black-and-white TV.

There was a lot to be done the next day, November 13, so I tried to get started early. My leg was stiff, the knee badly swollen, and I had a rash over most of my body, so I went to see Dr. Moreno, the GP, as soon as he was open. He gave me a tetanus shot and prescribed some antibiotics, anti-inflammatories, and an antihistamine. He also advised that I stay off the leg and get an X-ray as soon as possible to make sure the meniscus wasn't compromised.

From there I went to fill my prescriptions and then on to the customs office to try to do something about the truck's status. When you bring a vehicle into a country, they give you a permit for a temporary stay and note its presence on your visa. This action has often caused problems during my research because there are many times when we didn't want the vehicle to leave with the person who brought it in, and the failure to comply with your visa can have serious repercussions—like jail time. Despite the potential for disastrous consequences, I have exited Mexico at least three times without the vehicle that was supposed to accompany me. I always figured that my research permit, plausible explanations, and tears (if necessary) would help the customs officials see the wisdom in letting me slide, but in fact, they had never caught me. Each time it happened, some momentary distraction at exactly the right moment had allowed me to escape before they had time to stop me. Nevertheless, I thought it wisest to try to make some arrangements with customs in order to leave the truck in the country. I had had some earlier discussions with them back at the border when I brought the truck in as to what I should do. They had told me then that I could impound the vehicle in a customs lot and pay duty and a fee for them to keep it until my return. Now as I was making plans to head back to the States, I decided to go to the local customs office in Punta Gorda to see what I could work out.

The exchange was instructive. At the office, the customs guy played a little game with me. First, he valued the truck at $5,700 (an absurd figure) and calculated duty (which had to be paid to leave it in country) at $2,650. I laughed heartily at this and told him no thanks. So then he did some quick refiguring and came up with a duty payment of $750. That sounded doable given the hideous effort required to get it there in the first place. So I told him I would think about it. He then said that if I decided not to do it (pay the duty), I should not tell the customs people in Belize City what he had valued the truck at.

All right. On to the next events: get an X-ray at the clinic, pick up maps at the land office, call Dr. Belisle at the Forestry Office in Belmopan about the permits, check with our mechanic about the truck's broken front end, and so on, and so on.

The following day, November 14, we headed out to try to get into the forest north of the British military base at Salamanca. Some explanation is required, I suppose, for why such a place existed at the time. Above, I alluded to the fact that Guatemala had never recognized Belize as separate from its own nation. This issue, though unresolved, had been largely diplomatic until 1948, when Guatemala threatened to enforce its supposed sovereignty by invasion. The British responded by stationing two companies of the 2nd Battalion of the Gloucestershire Regiment in Belize. Nothing came of that particular threat, but things heated up again in 1972, resulting in a squadron of Harrier jets being stationed in the country in addition to artillery and infantry troops. The British military presence remained even after Belize achieved its independence in 1981 in recognition of the continued expressed intention of Guatemala to forcibly attach the country to itself. Things finally calmed down in the early 1990s, when, under pressure from the international community, Guatemala finally agreed to recognize Belize's independence. Most British forces left at that point, but some remained as "training battalions" up until 2011. In the meantime, Guatemala has continued to push its case through the international court, which has agreed to rule on it in 2022 or thereabouts.

If you look at a good satellite map of the terrain and existing roads, you will see that there is really only one good invasion route from Guatemala into Belize—namely, along the Belmopan–San Ignacio Road to the Guatemalan border. However, you will also see that the area along the border west of Punta Gorda, where access routes shrouded in forest are invisible to aerial surveillance, offers a relatively easy opportunity for harassing incursion, an activity engaged in fairly often by the Guatemalan military at the time, which explains the presence of the Brits at Salamanca, eleven miles east of the Guatemalan border.

We arrived at the base in the middle of the afternoon and were cordially welcomed by Maj. Andrew Duncan, commanding officer of the unit in residence at the time, the 1st Battalion of the 2nd King Edward VII's Own Gurkha Rifles (the Sirmoor Rifles; "Better to Die Than Be a Coward"), an elite group with a storied pedigree. When I stated my request, he invited us up to the officer's mess—an airy, screened-in second-floor veranda—for a more leisurely account with an iced Coke. After hearing a description of our project, he readily granted permission to set up our nets and camp in their forest, only warning us to watch out for armed Guatemalan commandos at night, a circumstance already mentioned to us by our contacts in the Q'eqchi' Mayan village of San Antonio.

After departing the base, we headed north a half mile or so along a barely passable track to a spot where we could park. We grabbed our gear and cut our way through some second-growth edge and into the forest to set up camp. The forest was obviously disturbed, probably high-graded within the past fifty years. As elsewhere, diurnal mammals and large birds were missing.

On November 15, 1989, we heard tayras, kinkajous, and owls in the night, a further indication that the lack of large birds and diurnal mammals wasn't a problem with the habitat. The disturbed rainforest was apparently fine so far as the usual animal denizens were concerned, but hunters had evidently cleared out anything worth a bullet among the day-foraging species. We started setting up our grid of thirteen nets at dawn and finished by about two in the afternoon, working mostly in a steady downpour—an unconscionable amount of time for the job, really, but my leg gave me a lot of trouble. I kept tripping and falling, which caused the knee to throb and the wound to bleed. It was just seeping, but worrisome. Plus the rash, which covered most of my body by this point, was bothersome. So as soon as the nets were in place, we took off on the two-hour drive to Punta Gorda to see if I could catch Dr. Moreno. I did. He didn't know what more he could do about the rash except change my antibiotic from penicillin to erythromycin in case it was an allergic reaction. As for the leg, he said he could not imagine a worse activity for it than stumbling around in rainforest in the rain. He was pretty severe. "It's your life," he said, "but if you break open that artery again, it might be a lot shorter than you had hoped." Overly dramatic perhaps, but an effective argument in my weakened and sleep-deprived state. I decided that we would end this field season and leave for Belmopan the next day.

With the final decision made, we wrapped things up at Salamanca and Punta Gorda on the morning of the sixteenth and headed out on the eight-hour drive (assuming no breakdowns) for Belmopan. The truck did quit at the top of the Dangriga-Belmopan pass, some electrical problem, but we managed to get it going again, arriving at our destination at 9:30 that evening. The truck quit again, but we were in the hotel parking lot, so we could at least get some sleep in a good bed before trying our luck.

We got the truck going again the next morning and finished the last leg of the trip into Belize City. I checked in with customs and placed the truck in their impoundment until our return in February. From there we got a taxi to the international airport and a flight out back to the States. Whew.

Our next Belizean field season began on February 5, 1990, when Bill McShea, my new colleague from the Smithsonian Conservation and Research Center, and I flew early out of Dulles Airport, arriving in the midafternoon at Ladyville. We took a cab from there to the customs vehicle impoundment in Belize City to arrange for the truck's release. They said I could have it at 8:00 a.m. tomorrow, so we left from there and went to the Chateau Hotel in town for the night. The next morning, we went out to pick up the truck. Amazingly, it started. The left front tire was flat, but a half hour of work with a bicycle pump got it ready to go.

We went from there downtown to check on what maps I could find at the post office. There wasn't much on-the-street parking nearby, but I managed to squeeze the truck in next to the building with just a little bit sticking out into the crosswalk. As I was getting out, a sturdy Black woman in T-shirt and shorts crossed from the far curb and walked up to address Bill, who was seated on the driver's side with the window down. I did not hear the exchange, but I saw Bill's expression go from a welcoming smile to one of alarm. I paused and leaned back into the truck. "What did she want?" I asked. Bill answered that she had told him that we needed to get our "motherfuckin' White asses" off the crosswalk.

Ooooookay. So I moved the truck.

On arrival in Belmopan, I went to the Forestry Office to deliver copies of a published professional paper we had written on our work (to establish our bona fides) along with the various letters of support that Dr. Belisle had wanted for our permits. We then picked up the field equipment that we had left in his care the previous November and headed south for San Antonio at about noon. There were truck problems along the way, of course (a hole in the radiator, an unidentified grinding sound from the front end), but we pulled into the village without too many delays later that evening.

We had stayed at the Heston brother's hotel, the St. Charles Inn, in Punta Gorda during our last field season, but the hour-long commute, one way, from there to the closest real forest of any kind was too time-consuming, and I had not enjoyed our nights camping. Fortunately, and to my great surprise, I had discovered a boarding house right in "downtown" San Antonio late in our previous November visit, Mr. Bol's extravagantly named Hilltop Hotel.

This establishment, operated by a Q'eqchi' Mayan couple, Mr. and Mrs. Bol, consisted of a downstairs kitchen and dining area, accommodating eight or ten people (including the cook), and three large upstairs rooms with three or four single futons on wood-slat cots in each. There was an outhouse for calls of nature and a small attached room at the back of the house with a basin and showerhead for cleaning up. In addition, Mrs. Bol and her young maids were perfectly happy to prepare simple morning and evening meals as well as to take care of our laundry for reasonable fees.

I assumed that we would have the Hilltop Hotel to ourselves and so was surprised to find two guests seated in Mrs. Bol's kitchen when we arrived that evening: a man in his forties with a comely younger woman, perhaps in her early thirties. I try to avoid fellow foreigners when working in the field. They are usually needy, lonely, bored, and desirous of a willing listener to lengthy accounts of the sights they have seen and whatever current plight they are in. In this case, however, we were caught. Mrs. Bol's kitchen is not large, so we couldn't help being regaled—at least for the duration of the meal.

You may have seen the classic Bud Dry beer commercial "Why Is a Good Man So Hard to Find?" If not, I advise you to check it out on YouTube. It's pretty funny—unless you happen to be trapped with one of the guys depicted. The commercial is set up with the camera presenting the woman's view of her date, who is seated across from her at a table for two in a nice restaurant. Each of these men represents an obvious type and is shown addressing his date with conversation exemplary of their class:

> The Lothario: Lots of women find my looks intimidating. Do you?
> The Milksop: My mother makes the best brisket.
> The Tycoon (talking expansively on the phone): Sorry. I have to take this call.
> The Gorilla (reaching across with his fork for her plate): You gonna eat that?

And so on.

The type we were cornered by was The Adventurer (per the commercial, "There I was! There I was! There I was! In. The. *Congooo!*"). Our representative of this genre explained that he was a professional photographer (shades of *Bridges of Madison County*), currently not on assignment, taking his lady friend (probable funder of the trip) to show her some of nature's most remote, spectacularly beautiful, and exciting wild places. Now, I have to admit that the serene (except when invaded by Guatemalan soldiers) village of San Antonio had its charms, but none of the adjectives used by our dinner mate came to mind when considering it. Anyhow, after filling us in on his experiences and plans through the meal, we excused ourselves. He said that he would really like to go with us to check our nets in the morning. Dinner was bad enough, but the prospect of having this gentleman tagging along, trying to make sure we were always within earshot of his prolix tales, was an unhappy prospect for me. I told him we breakfasted at 5:00 and would be out in the woods by 5:30, so it just wouldn't work. He said no problem. In the unlikely event that we had left before his awakening, he would find us along the road.

The next morning, we were up and out before he could catch us, and I thought we were safe. Not so fast, my friend. To my great surprise, we heard him hallooing for us near where the truck was parked along the road at the Salamanca site. We hid, and eventually he went away.

Later that morning, while Bill and I were checking our mist nets for captures, I heard a loud *krump* a bit off to the west of us. Now, as an artillery liaison officer in Vietnam, I had heard this sound nearly every day, which I took to be the explosion of a 105 round, so I grabbed Bill, moved quickly out to the road, and drove over to talk to the Brits. On arrival, I tracked down the commanding officer, Major Andrews (Major Duncan had rotated out), and asked him what was going on. He was surprised to see me. He had no idea that we were out there. The area where we were working was a "free fire zone," a term I was intimately familiar with. A free fire zone is an area whose boundaries have been negotiated with the local government such that artillery can be fired there whenever the military wants to do so without notifying anybody. Everybody knows about this, so if you are out there and a round happens to land on you, that's your problem. No legal recourse. We were the only ones ignorant of the arrangement.

Over the next few days, Bill and I were able to net in forested grids at the site north of Salamanca and at Alf Koon's place, catching and/or sighting a lot of migrants, including Great Crested Flycatcher, Summer Tanager, White-eyed Vireo, Eastern Wood-Pewee, Yellow-bellied Flycatcher, Gray Catbird, Wood Thrush, Kentucky Warbler, Hooded Warbler, and Black-and-white Warbler. We also caught a bunch of bats of several species, which were Bill's main interest.

On traveling to one of our pasture netting sites, I happened to notice a long row of beehives. This finding I found interesting, and so I drove to the nearest residence to see if I could locate someone who could tell me about it. No problem. In about five minutes, I managed to find one of the keepers. My interest stemmed from the fact that "killer bees," an African form of the honeybee, had been accidentally released in Brazil in 1956. This type is much more aggressive than its domestic cousins, attacking anything approaching within fifteen feet or so of a hive. Also, the sting of one bee incites all neighboring bees, apparently through the release of pheromones, to swarm and attack the potential intruder. If you have the bad luck to be allergic to the venom, an attack of this magnitude is likely to be quickly fatal—hence, "killer bees." The bees released in 1956 found the New World to be quite hospitable. They hybridized readily with local bees, and unfortunately, the hybrids possessed the aggressive nature of the African form. Distribution spread steadily northward during the years after their Brazilian release, reaching northern Brazil by 1971, Colombia by 1977, and southern Mexico by 1982. This movement terrified some people in the United States, an emotion fed by the hysterical (in both senses) 1978 movie *The Swarm*, in which killer bees attack Houston after slaughtering a bunch of Texans elsewhere in the state (Africanized bees have been in Texas at least since 1990, including at the Welder Refuge). Aside from the mania, I wondered how Africanized bees had affected the honey business. My interlocutor said that killer bees had arrived in their hives about a decade ago. He admitted that they had had to make some adjustments in how to handle the hives for harvesting the honey, but once those had been made, they had no problems. The hybrids were more aggressive, it is true, and they didn't produce quite as much honey per unit. However, they seemed to be hardier than their European relatives and less susceptible to disease.

I had other projects underway back up in Virginia and couldn't stay past the eleventh, so George arrived on February 9 to take over for me. He and Bill continued netting and censusing at our reconnoitered sites up until February 18, then flew back to the States, leaving the truck parked at the Toledo District Forestry Office just outside Punta Gorda and our gear with the chief of the Forestry Office, Mr. Edwards.

My next trip to Belize was planned for November 1990. I left Virginia on November 5 accompanied by Tim Fadness, a young undergraduate volunteer from Minnesota. We made the trip—Dulles to Miami; Miami to Ladyville (Belize International Airport); Ladyville to Belize Municipal Airport (Belize City); Belize Municipal Airport to Punta Gorda—with no significant problems, staying the night with the Wagners at their St. Charles Inn. The next day, November 6, was spent trying to get our truck and equipment from Mr. Edwards at the Forestry Office. He wasn't there, but we convinced his subordinates to at least let us take the truck. They didn't know where he had put our equipment. Eventually, we tracked him down at home and got our nets and other necessary materials for carrying out our fieldwork. On November 7, we set off for San Antonio for our stay at the Bols' Hilltop Hotel.

Anyone attempting to do fieldwork anywhere without collaboration with local people is a fool. No number of visits can provide the knowledge of the land that these folks have. I knew this, of course, from my years in Texas and Mexico, but circumstances regarding the way in which the project developed through Chan and his team, and the fact that Belize had no native field biologists of whom I was aware, delayed the development of this type of necessary relationship, and it was not until this third field season in November 1990 that I was able to get a Belizean onto our team. The contact came through our landlords, the Bols, and their employees, who had become our friends. We did shopping for them in Punta Gorda when possible and brought them a few special items from the States. Pretty much everyone is related to everyone else in San Antonio in one way or another, and so at breakfast when I was complaining to Mrs. Bol's maid, Isabel, about the lack of a person with local knowledge on our team, she mentioned that her cousin Patricio Choc might be interested in helping us out. "Fantastic," I said. "Ask him if he could meet us when we come back in for supper at six this evening."

And so there was Patricio at supper. I had prepared for the meeting by asking Mr. Bol what he thought a fair wage would be, and he suggested $15 a day. I offered Patricio that, which he accepted, and we had our local person on the crew. He was outstanding, working for us every day whenever any of us were in-country for the next two field seasons until our final departure in the spring of 1991. He quickly learned the few skills required for our work: taking birds out of nets; keeping notes on date, time, and species; and banding and recording the data. He also learned that we wanted him to tell us anything that came to mind concerning land use. This information was invaluable. For instance, I had been grossly overestimating the age after clearing for second-growth areas, data he was able to correct with precision.

Mr. Bol, in addition to his guesthouse, also ran a small general store. Behind the display area for this establishment, there was a room with folding wooden chairs that seated about twenty. A couple of evenings after our November 7 arrival, Mr. Bol invited

Figure 24-5. Patricio Choc with a White-collared Manakin (tropical resident) in his right hand and a Gray Catbird (Nearctic migrant) in his left.

us to join him and other village elders in this back room for their regular Friday Bible study. We demurred, pleading exhaustion. Two nights later, he invited us to join the elders again for their regular Sunday evening viewing of "adult" movies. We passed on the invitation again, believing that, as is the case of the Bible study, our fatigue would preclude giving the event the attention it deserved.

I have said that the village of San Antonio was a quiet place. I meant this in the sense of activity. As in the majority of farming communities, most everyone shut down and went to bed after sunset, leaving the darkened lanes to the occasional Guatemalan squad. In another sense, however, it was not quiet. All my life, starting I guess with Mother Goose and Chanticleer, I had heard that roosters crowed at dawn. The San Antonio roosters weren't raised right or hadn't gotten the memo. They crowed off and on all day, and when night fell, they really cranked it up, keeping a steady chorus throughout the hours of darkness. Furthermore, they were joined by all the dogs and burros in town, maintaining a perfect bedlam from dusk to dawn. For most of my helpers and me, the chorus was no more than white noise—not exactly restful, but not enough to keep you awake after a long day under the truck or cutting net lanes. Some, however, found the din maddening. Like Tim and George.

During all of my previous visits to Belize, contrary to my experiences in Mexico, I had never encountered or even seen a government official of any kind outside of his office despite hundreds of miles traveled from one end of the country to the other—not customs, military, forest service, constabulary, nada. That changed on February 12, 1991. I was driving along the dirt track from San Antonio to Blue Creek when I was stopped by the police. Now, Blue Creek is a Q'eqchi' Mayan village of perhaps 150 souls. They had no police, of course. This pair derived from the Punta Gorda station, and the possibility that this meeting was a simple result of a routine patrol was likely to be near zero. Someone, probably one of the larger landowners, obviously was uncomfortable with our activities. Anyhow, the cops gave no reason for the stop. The facts of my disreputable appearance and foreign derivation were apparently deemed sufficient. They looked over my permits, visa, and car papers with obvious skepticism but, finding nothing immediately illegal, decided to let me go—for the time being. Five days later, two of my team were stopped by police again, and this time they had thought of something that might be amiss. Where was our Belizean insurance? Hah! We had none. They advised that we had to have the insurance and that we should purchase it immediately and then report to the police station in Punta Gorda with proof of purchase. The next day, we bought the required insurance for $50 and repaired to the station. The duty officer was completely perplexed, knowing nothing about our encounter or required immediate appearance. Nevertheless, he took our names and confirmed that we had completed the required task, and that was that. No more police problems.

I hate to speak ill of the dead, especially one that served me and my team members for several years under a variety of exceptionally difficult circumstances, but I must make some mention of the Blazer's final weeks of life in southern Belize. The various maladies included the following:

1. Dead battery (multiple times; see "Dead alternator" below)
2. Dead starter
3. Dead alternator (brushes bad; on order for weeks)
4. Broken positive battery cable (grounded)
5. Broken fuel filter
6. Broken fuel line
7. Overheating as a result of several different problems (low oil, radiator leak, lack of coolant, broken radiator hose, etc.)
8. Flat tire
9. Ruined brake pads
10. Corroded brake cylinder (causing hub overheating—and fire, of course, if not stopped immediately)
11. Detached undercarriage brace (causing the engine drive train to drop down unless tied up with something)

The brake cylinder problem had occurred before in Catemaco back in September 1983, causing the right rear wheel to catch fire. The quick use of my pants had rescued it from complete immolation at that time. Some days after my departure during our final Belize season, my team wasn't so lucky. The Blazer caught fire and, like a Viking warrior, burned to a shell. One could argue that it is tempting fate to bestow unfortunate names, like Dolores, as they are likely to influence destiny. Certainly, it did not turn out well for our Blazer.

On the afternoon of February 22, 1991, I left the Toledo District for the last time, turning over the remaining month of sampling to my research squad and taking the taxi Cessna out of Punta Gorda to the Belize Municipal Airport.

My flight home was not scheduled to depart until the next morning, so I had to decide where to pass the night. The municipal airport is about a mile from the downtown hotels, which didn't seem so far, so I headed off, lugging my two suitcases (sans wheels). I have mentioned a couple of my earlier encounters with Belize City residents in this account. Suffice it to say the number of pimps, pushers, pirates, and other like-minded entrepreneurs poised to pounce on visitors seemed disproportionate in my experience, so I was not surprised to be accosted by a young man within a block of setting out. The principal language spoken at that time in the town was a Caribbean patois intelligible only to residents. However, most folks spoke perfectly clear English with a Marleyesque lilt in need, and it was in this fashion that my new friend addressed me. He wanted me to know that he recognized me as a traveler needing assistance, which he was perfectly capable of providing, beginning with, but certainly not limited to, carrying my luggage. I told him that I didn't need his help for the specific job offered or any ancillary offices. He insisted, and so we continued along for the next block. Finally, I realized that he was going to accompany me all the way to my hotel whether I let him carry the bags or not, so I might as well let him do what he so ardently desired. I asked how much he would charge. He said $10 Belize ($5 US). I laughed and said I could have taken a cab for that. He said yes, but I hadn't, and we went back and forth until, after another block, agreeing on a price of $4 Belize. He then hoisted the bags with ease, and we continued in conversation the rest of the way to the hotel, with my interlocutor in continuous sales mode—lots of things to do in Belize City at night if you know the right people, and he knew them all, and so on. Arriving at our destination, I gave my new friend the promised $4 Belize for the service and $6 Belize for the entertainment. The hotel's commission-aire examined my garb of worn (but clean, thanks to Mrs. Bol) field clothes skeptically before allowing entry (as many a doorman in many a country had before him). I entered, checked in, spent the night in luxury, and flew out the next morning for home, bringing my Belizean adventures to a close.

Our purpose in going to Belize was similar to that for our work in Costa Rica. We wanted to find out what habitats wintering migrants were using and to see if we could use remote sensing with satellites to determine how much of those habitats remained. After the completion of our Belize study, we went back to the Tuxtla region of southern Veracruz, Mexico, and spent several field seasons there with the same ideas in mind. The results from Costa Rica and Veracruz were comparable. A community of migrants wintered commonly in older tropical forest, of which little remained in either country. Many of the same species could be found in younger second growth and even pasture if some trees were present, but at much lower density. The Tuxtla data were especially striking. Satellite information showed that about 10 percent of the region remained forested.

However, almost all of this remaining forest was in the rugged volcanic mountains above 1,500 feet, elevations too high for many of the migrant species, like the Wood Thrush and Hooded Warbler. Less than 1 percent of the forest below 1,500 feet, suitable for these migrants to winter in, remained.

Belize, however, was different. Nearly all of the forest that we sampled had been disturbed at some point within the past fifty years by high-grading—that is, by removal of valuable timber such as mahogany. This activity did not appear to affect either the migrant or resident bird communities, which appeared to be comparable in composition whether the forest was disturbed or not. We found this to be true in Costa Rica and Veracruz as well. The big difference between these countries and Belize was in the prevalence of older second growth. Areas where forest has been completely removed in the part of Belize where we were working evidently was often allowed to grow into older second growth and young forest, and these habitats held migrants, although how valuable they were in terms of supporting populations through the winter was not obvious. The reasons for this difference with Belize as opposed to Mexico and Costa Rica, I believe, had to do with land use policy and population density. In Mexico, for instance, the *ejido* system implemented in the 1930s provided for the free distribution of land to petitioners, which, along with a high population of landless peasants, resulted in near complete clearing of arable land for subsistence farming. This land was kept cleared once trees were gone, either for pasture or for farming. The farming practice used was "slash and burn," or "swidden," in which the forest is cut and burned, then farmed until the soil is exhausted—usually a year or two depending on rainfall, soil type, and slope—then allowed to go fallow for a period of two or three years before cutting and burning again. Therefore, older second growth is uncommon or rare except along roadsides or streambeds. Not so in southern Belize, where the population is low away from the coast.

We had done a great deal of work on migrant use of forest and second growth in Mexico, as I have described in earlier chapters. We had found that migrants wintering in forest remained there throughout the winter, whereas those using young second growth did not. They seemed to be continually on the move. We suspected that those migrants that were moving more than members of the same species in forest suffered higher mortality and were able to document that this was, in fact, the case for at least one of the species, the Wood Thrush, through an expensive and arduous long-term study using airplanes to track and compare survival outcomes for resident territorial birds in forest as compared with wandering, nonterritorial birds in second growth (see chapter 23).

All told, I spent four field seasons in southern Belize: November 1989, February 1990, November 1990, and February 1991. I had some help from various people during these endeavors, as I have documented above, but the funding, design, logistics, and execution of the project were entirely my responsibility. Therefore, it was with dismay that I received a draft of a technical paper based on the Belize data from the field assistant, Joe Brown (a pseudonym), whom I had hired to continue the work in my absence during the final two field seasons.

That Joe had done an excellent job under difficult circumstances in carrying out my study design is without question. Nevertheless, an assistant is an assistant. Performing the duties for which they were hired does not entitle them to take credit for the research in the marketplace of ideas represented by professional publications. Despite this obvious ethical problem, my options for dissent were limited in my view by the fact that Joe had enlisted a professional consultant for the statistical analysis, and my coprincipal investigator, George Powell, had already read and approved the draft. Both men had now invested in the paper, and their names were on as coauthors. My name was there too, although I had never seen even a preliminary version despite the fact that the draft in my hands was obviously now in its final stages before submission to an appropriate journal. But this was not the worst of it. Unfortunately, the paper represented a textbook example of the misuse of statistics to reach questionable conclusions, as I explain below.

Data collected on habitat distribution and amounts and its use by migrants during the first two field seasons in the Belize Toledo District showed pretty much the same thing as our Costa Rica work. Several species of migrants could be found in lowland forest as well as in second growth, but there was hardly any forest left—indeed, there was no primeval forest whatsoever below a thousand feet that we could find. Our sampling procedures, however, gave no hint of the relative value of forest versus second-growth habitat for migrants. As I have detailed in previous chapters, we were able to get at this question in Veracruz with immense difficulty, using years of radio-tracking of Wood Thrushes in primeval rainforest and scrubby second-growth habitats, which had shown that mortality rates for members of this species were much higher for birds using second growth than for those using rainforest. In thinking about our Belize findings during the preparations for the final two seasons, November 1990 and February 1991, it occurred to me that we might get an inkling of relative value by establishing ten small (2.5-acre) netting grids of thirteen nets each in second growth and ten grids in forest; then run each grid for a thousand net hours in both habitats, once in November and December, banding and releasing all migrants; then return to each grid in February and net for a thousand hours again. We would then compare the recapture rates found for migrant species at the forest sites as opposed to those in second-growth sites. If sample sizes were large enough, we might be able to test persistence as a stand-in for value for migrants in forest versus second growth.

After the project had been completed, I examined the data and could see right away that we just didn't have enough to say much on the question. Our sampling periods had been too brief and our captures and recaptures too few in both forest and second-growth habitats to make much sense of them. Joe evidently disagreed, although he said nothing to me. He took the data to a statistician and asked that he calculate "survivorship" in the two different habitat types based on migrant recapture rates. Now, there are two problems with this approach. This type of analysis is standard in many wildlife studies, but the underlying assumption is that an animal that disappears is dead—hence the use of the term *survivorship*. We already knew from years of work in Veracruz that this was not the case for migrants. Migrants in forest simply moved around less than members of the same species in second growth. Naturally, this moving around leads to lower recapture rates, but it is not the same as death. The second problem was one of probable type 2 sampling error in our data. A type 2 sampling error occurs when the sample size is too small and the variability too large to reach a meaningful result, which appeared to me to be the case for our data. My assistant thought otherwise. The paper's conclusions read as follows: "We estimated overwinter survival and capture probabilities of Wood Thrush (*Hylochichla mustelina*), Ovenbird (*Seiurus aurocapillus*), Hooded Warbler (*Wilsonia citrina*), and Kentucky Warbler (*Oporornis formosus*) inhabiting two common tropical habitat types, mature and early-successional forest. Our results suggest that large differences (for example, ratio of survival rates (γ) ≤ 0.85) in overwinter survival between these habitats do not exist for any of these species."

My heart sank when I read this. The data, of course, showed no such thing. Recapture rates are not survival rates, and insufficient sample sizes did not allow a valid comparison of recapture rates between the two habitat types for any of the species mentioned. The most that could be said would be that our data, weak as they were, didn't show much difference in recapture rates between the two habitat types. I immediately contacted Joe with my objections. He simply said that he, as senior author, and the other coauthors disagreed.

I had no real options, as he knew, so I told him to take my name off the paper, which he did. And so this piece of our Belize work came out in a major journal published in 1995.

The above diatribe may sound a lot like whining, which I suppose it is. After all, it is not at all uncommon for students or technicians on a project funded and planned by others to attempt to publish the data on their own. It has happened to me on three other occasions. What made this one so painful was that a corps of researchers led by Mario and me had spent five years and several thousand dollars attempting to find a valid measure of survivorship in forest versus second growth in one species, the Wood Thrush. To have those findings questioned in a prominent journal article based on such weak reasoning using paltry data collected in my own study left a bitter taste.

Nevertheless, and despite my various vehicular and personnel issues, my last words on Belize must be overwhelmingly positive. It is a beautiful country with spectacular beaches, glorious tropical flora and fauna, and a fascinating mix of cultures, most of whom are welcoming. Furthermore, it is readily accessible from the States, relatively inexpensive, and English is the official language!

THE SHUAR

Nada se obtiene sin esfuerzo.
(Nothing can be gained without force.)

—Maxim of the Shuar

A conclave of indigenous peoples from around the world took place at the Smithsonian in the summer of 1991. As part of that meeting, excursions to our Conservation and Research Center were offered to those participants who might want to get out of DC into the wilds of the Blue Ridge Mountains and see some of the animals we kept for captive-breeding programs, like red pandas, clouded leopards, and maned wolves.

I led one of these tours and, using my primitive Spanish, got into a discussion with one of the visitors about our research work on migratory birds in Middle America. This gentleman, Don Miguel Puwandir, happened to be president of the Shuar Federation, a group whose homeland was in the far western portion of the Amazon basin, two thousand miles from the mouth of that great river near the cis-Andean base of the mountains in Ecuador. He expressed evident fascination with our bird studies, and I, always on the lookout for new opportunities for migrant work in new parts of the Neotropics, wondered if his people might be willing to collaborate with us if we were able to find funding. He said that he would certainly consider it, and we parted in mutual satisfaction after exchanging contact information.

There was a sublime serendipity about this chance encounter because, as luck would have it, the fourth international meeting of the NOS was scheduled for November 3–9 in Quito, Ecuador. I was an officer of this society (Mario was president), and I had planned on attending anyway even before the possibility that this event might provide an opportunity to further develop a relationship with the Shuar that might lead to an investigation of intratropical and southern migration. Accordingly, I sent a letter to Don Miguel, care of the federation's office in Quito, asking if it might be possible to take the

Figure 25-1. Shuar emblem showing a warrior holding aloft an enemy's shrunken head.

opportunity of my trip to visit the Shuar. He wrote back immediately to say that they would be delighted to have me visit. Specific arrangements could be finalized in Quito after I arrived in-country.

I scrambled around then, trying to find the funds to let me stay an extra week or so in Ecuador and make the long trip across the mountains to Sucúa, hometown of the Shuar Federation. I also tried to enlist Mario and George Powell to accompany me. Mario had other commitments and couldn't make it, but George agreed to come along.

Prior to my departure, I did a little background research on the Shuar. They are an indigenous group living at the headwaters of the Marañón River in the tropical rain-forests of the far western portion of the Amazon basin at the foot of the Andes in Ecuador and Peru.

Like the Comanches and the Scots, and as indicated by the material shown at the beginning of this chapter, the Shuar have a well-earned reputation for ferocity. In the "old days" (a hundred years ago? fifty years ago? yesterday?), adulthood in males was achieved by killing an enemy, taking and preserving his head as a token of having passed that crucial initiation into society. In addition, Shuar males did not believe in natural death. Every adult male died either as a result of being killed in battle or by the secret attack of an enemy. But there was more to the Shuar than savagery. Like the Jews and the Spartans, they are an ethnocultural entity that has maintained itself through a

Figure 25-2. Shuarland in the Amazon basin.

regimen of fierce independence and recognition of their own specialness through not only continual war but education as well.

The Shuar reputation for independence came early. Although perfectly happy to trade with other cultures, they violently resisted any attempt at colonization or conquest. In 1527, they defeated an Inca army led by Capac, and in 1599, they ousted forces of the Ecuadorean government attempting to coerce taxation. Today, although they recognize themselves as a part of the country of Ecuador, they are in fact a federated entity, maintaining their own civil authority and educational services. Shuar males participate as volunteers in the Ecuadorean army, if they wish, where they constitute their own elite force, the Iwia ("forest demons," in the Shuar tongue).

With this basic introduction in mind, I set out. I flew into Quito on November 2, rented a little Suzuki at the airport, and headed off to the hotel where the bird meetings were being held. These were contentious, as always seemed to be the case. The main difficulty this time was that the man chosen to be the editor of the society's fledgling journal had failed to publish even one issue in the four years since his appointment. Obviously, he would have to be removed and replaced, an admittedly demanding task but one that was essential to the future of the organization.

Mario had found a replacement in Professor Doctor Karl (Charly) Schuchmann, curator of birds at the Alexander Koenig Museum in Bonn, Germany. Charly, arguably the world's leading expert on hummingbirds, had decades of experience in the new-world tropics. Mario decided to make the change official at a meeting of the executive council, at which I was present as treasurer. In addition to the other officers, the group included the present editor, Humberto Alvarez, a Colombian professor from Bogotá whose PhD was from Cornell, and Charly, his potential successor. When Mario announced the switch, Humberto was quick to take exception. In his usual soft-spoken tone, he made some singularly vituperative comments, largely directed at Mario's leadership. Nevertheless, the deed was done, and Charly justified Mario's confidence in him perfectly, producing four excellent journal issues a year for the next ten years until his retirement.

On the second day of the conference, I was scheduled to meet with the Shuar Federation *presidente* at his office downtown just off the *zócalo* at 1:30 p.m. to finalize the arrangements for our postconference visit. I left the hotel in the Suzuki a few minutes early to see if I could find a shop where I could buy some additional film. In this I was successful, but on returning to the parked jeep, I discovered that I had locked my keys inside. With five minutes to go and about a mile to travel to the Shuar headquarters, my options were limited. I picked up a large piece of pavement from the curb, waited until the cop standing on a corner about a block away was looking elsewhere, and smashed the window. I got in and drove off without any further excitement, arriving at the meeting on time. A week later when I turned the vehicle in, I explained to the agent that some accident had occurred resulting in the broken glass. He smiled, said "No problema," and explained that window replacement was not covered by the insurance and that he would be happy to receive $200 in payment. Oh well.

The meeting with the Shuar went well. We set a date and time when George and I would be at their offices in Sucúa, where we would be given a VIP tour of the Shuar Federation facilities and then flown for a visit to one of their member villages. We would be put up in their guesthouse for the night and given dinner that evening and breakfast the next morning. Fantastic.

I also took the opportunity during my stay in Quito to try to visit the in-laws of my cousin Annick (daughter of my dad's brother, George). She had married an Ecuadorian agronomist when both were working in Guatemala. His parents had an apartment downtown for which Annick had given me the address. I think it was probably a shock to have a gringo show up on their doorstep unannounced, but they took it with remarkably good grace, inviting me in for tea and cakes and some desultory conversation about mutual relatives.

George and I left Quito on the ninth to begin our adventure with the Shuar. It's a long drive, first south along the Andean cordillera for three hundred miles, then east a hundred miles and down several thousand feet. It took eight hours to get to the town of Gualaceo in the foothills, where we spent the night in a pleasant hotel.

Setting out the next morning after breakfast, we arrived at the home base of the Shuar Federation in Sucúa on time for our appointment with the vice president. He sat us down in his comfortable office, gave us some coffee and cakes, and proceeded to tell us about his people. He began by explaining that the Shuar reputation for rigorous enforcement of their autonomy had allowed them to maintain their ancient way of life well into the twentieth century. However, by the 1950s, increasing outside encroachment by settlers from the trans-Andean portions of Ecuador brought them into a more confrontational mode. They therefore began negotiations with the Ecuadorean government, according to which the Shuar would agree to become a federated unit of the country, paying taxes and participating in other national responsibilities such as defense, in return for which the government would provide assistance in controlling immigration onto Shuar lands. This arrangement was formalized in 1964. As a result, although the Shuar are technically a part of the country of Ecuador, they are mainly self-governed.

Today there are roughly forty thousand Shuar, living principally in a few towns at the base of the Andes and about two hundred villages in the Amazon basin. For the most part, these villages are connected only by footpaths, so the primary linkage between them and other population centers of the federation is by air. Each village has its own runway.

The home base for the federation is in the town of Sucúa, consisting of administrative offices, library, air traffic control tower, hangar, radio station and broadcast antenna, menagerie, maintenance offices, shops, and garages. The governance of the Shuar is democratically organized, with popularly elected officers for each community and the nation as a whole.

With this introduction completed, the vice president took us on a tour of the federation campus, showing us all of the facilities mentioned above. During the inspection, the VP told us a couple of enlightening stories, probably apocryphal, about his community and its relationships with the wider world. Both involved the Dominicans, who spent a hundred years or so attempting to Christianize them. The first had to do with the exciting event of the opening of the first road passable for motor vehicles to their lands. They invited the priests to a major celebration to be held on a given date. The priests drove in, and the celebration took place with revelry well into the night. The next morning, the vehicles in which the priests had arrived were nowhere to be found. They formed the nucleus of the Shuar motor fleet.

Another story had to do with flight. Few Shuar villages can be reached by road, and so the Dominicans encouraged the Shuar to build landing strips next to each major village so that they could be reached by air. This the Shuar were happy to do, and in fact, they organized permanent work parties charged with the construction and long-term maintenance of each landing strip. The priesthood owned the planes, and priests were trained as pilots. The Shuar requested and were given instruction in flight. After all, you never know when a pilot might be hurt or ill and need a stand-in. However, not long after receiving training, a Dominican pilot flew into a remote village, spent the night, and awoke to find the plane missing! Like the trucks, it formed the first member of the Shuar squadron.

These stories—parables, really—are meant to give you some insight into Shuar culture. As a group, they are glad to associate with, trade with, and learn from outsiders. Nevertheless, no one should confuse such hospitality with any form of compromise of their independence. That they will do whatever is necessary to maintain.

The tour of the Shuar headquarters made two things quite clear. First, they place a heavy emphasis on education. Their library is extensive, and they have commissioned fifty-two textbooks (as of 1991) by Shuar scholars on all principal aspects of life: history, language, culture, and so forth. These books are distributed to all Shuar schoolchildren at the appropriate age for their instruction, and they receive classes in them daily from teachers in schoolrooms in the various villages and by radio from the Sucúa central station. This station also delivers information and entertainment for families in general throughout the nation.

The second striking aspect of Shuar life conveyed by the tour is the centrality of spirituality. While many Shuar are accepting of Christianity as inculcated by various missionary groups over the past few hundred years, most Shuar, particularly those still living in remote communities, are deeply reverent in a more inclusive way. They seem to have no problem with a "great spirit" as represented by the Christian God, but they know perfectly well that forest spirits play critical roles in their daily lives. Belief in the importance of these entities is, I believe, one purpose of their Sucúa menagerie. Many Shuar are unable to understand or even recognize these spirits because their lives are no longer spent in the forest. The menagerie is a place where those who live in towns or on farms or ranches can observe and be taught the importance of these animals in their spiritual lives.

I found it interesting to discover that the Shuar consider the term *indigenous person* to be as much of a slur as *Indian*. They understand that such labels just mean "ignorant savage" and reject all inclusive terms of reference of this type as meaningless at best. For them, it is as ridiculous to call a Shuar *indigenous* as to call the descendants of Europeans *invaders* or *colonists*. Similarly, they believe it is absurd to refer to the people of Middle and South America as *Hispanics* or *Latins* simply because Spanish is the official language of their countries. These peoples share no common culture or ethnicity; they are all different. It would make as much sense to call the people of the several countries of Africa that have English as the official language *Anglos*. Absurd. Such shorthand speaks of breathtaking ignorance.

Another interesting aspect regarding the Shuar is that while recognizing that outsiders may think negatively of them, they really don't seem to pay much attention to such opinions. In fact, they include in their curriculum a book written by a Dominican priest, Enrique Vacas Galindo, OP, which, while describing the culture and community as it existed in the late 1800s, calls the Shuar "savages" (*salvajes*). They don't care. It doesn't bother them. They think his writings have value in understanding their own

Figure 25-3. A Salvin's Curassow in the Shuar menagerie.

Figure 25-4. A Shuar village classroom.

history despite the unfavorable slant from which it is presented, and so they teach his book to their children.

After completing the tour, the VP put us on one of their planes, and we flew out over the vast Amazonian rainforest to one of their villages. There we were greeted by the schoolteacher and his wife and taken to the schoolroom to meet some of the village elders and partake of some refreshment (a milky, slightly effervescent drink presented in a half-nut shell about the size of a small cup). The teacher made us welcome with a short introductory speech, to which it was obvious I should make some response, which I did in lame, halting Spanish, a second language for all of my listeners.

After that, the teacher's wife, an unforgettable person, took us on a short tour of the village. She was in her late forties, I would guess, a woman of medium height and build with long black hair held back with a clasp. She was dressed in a light, colorful, cotton frock and walked barefoot. The most striking aspect of her otherwise normal twentieth-century rural female appearance was her face, which was finely tattooed with small black dots in the pattern of a flower centered on her mouth, with petals spreading across her cheeks, chin, and nose. Obviously, she was raised at a time when the ancient traditions still held sway. Nevertheless, there was nothing in her behavior or conversation that indicated anything out of the ordinary. In fact, she was open, smiling, and chatting throughout our visit, telling us all about her sons, of which she had three: one was still living in the village with his wife and family, one was living and working in Quito, and one had just moved to Los Angeles. Wow, what a world!

So that pretty much completes my experiences with the Shuar, obviously one of the world's most remarkable "ethnic" groups. Partnering with the Shuar, I tried to get some money from various sources to do research on migrants wintering in the Amazon rainforest and neighboring foothills of the Andes, but to no avail.

Interestingly (to me, at any rate), I was to read about them a couple of years later in an article in the *New Yorker*. The author's intent was to illuminate the rapacious behavior of oil companies exploring the Amazon basin and, in particular, their despicable treatment of the indigenous peoples occupying prospective drilling areas. The import of the account was that while all companies were pretty much amoral in their attempts to exploit these peoples, some were way worse than others. Shell, I think, was the leading example of bad guys, while Chevron did better. Anyhow, one of the stories related had to do with tribes associated with the two competing companies. It seems that some tribesmen associated with Chevron were traversing a contested area in a truck when they were waylaid and slaughtered to a man by tribesmen associated with Shell. The murderers were, of course, Shuar, who, as intimated by the author, had a reputation for viciousness.

KISS MY GOLDEN CHEEKS

*The Biology and Politics of Endangered
Species Conservation*

It [the Golden-cheeked Warbler] was listed as Endangered . . . in 1990, because of continuing concerns over loss of habitat, caused primarily by urban expansion and land clearing for agricultural activity. . . . Conservation of the species remains a contentious issue through much of the hill country of central Texas.

—Clifton Ladd and Leila Gass, quoted in
P. G. Rodewald, ed., *Birds of the World*

The federal Endangered Species Act of 1973 is arguably one of the most important actions ever taken for the preservation of the United States' natural environments. The purpose of the act was to identify those organisms threatened with extinction, protect them from further decline, and implement programs to bring about the recovery of their populations. Thanks to the courts, the act has proven to have tremendous power in terms of habitat preservation and restoration because these most threatened species generally prove to be the "canaries in the coal mine." Saving them often results in saving the entire system on which they depend, along with all the other species with which they share that dependency—for example, the Spotted Owl and western old-growth forest, the snail darter minnow and the Little Tennessee River, and so on. For endangered species of migratory birds, however, the situation is a bit more complex.

In 1985, when I was working as a research scientist at the CKWRI in Kingsville, Texas, I was invited to join the newly formed Golden-Cheeked Warbler Recovery Team preliminary to the official listing of the species as endangered (which occurred in 1990).

The golden-cheek is a species of migratory bird that breeds in the scrubby oak-juniper of Central Texas.

Sharp declines in its numbers had been recorded during the previous thirty years and were attributed to the destruction of breeding habitat resulting from the rapid conversion of ranchland to residential and industrial uses in the vicinity of such cities as San Antonio, Austin, and San Marcos. Aside from myself, our group consisted of people either involved in the study of the bird in Texas or concerned with its conservation in the state as land managers or concerned citizens.

Carol Beardmore headed up the group. She had studied the bird for her master's degree, and it was her job as a USFWS employee to staff the team, organize the meetings, and ultimately, assisted by Dean Keddy-Hector, prepare and publish the *Golden-Cheeked Warbler Recovery Plan*, which was completed in 1992 (currently available online).

My involvement in this effort was somewhat anomalous. Although I had done extensive work on migratory birds in Texas and Mexico, none had involved Golden-cheeked Warblers. In any event, I saw my inclusion, which I attributed to Carol, as a nod to the fact that golden-cheeks spent only four months in Texas: March through June. The rest of their annual cycle was spent somewhere in the Central American highlands, so far as was known.

Not surprisingly, then, my offering, quoted below, amounted to little more than a single paragraph in the eighty-eight-page plan:

> 1.310—Determine the current distribution and availability of habitat in the winter range and migration corridor. Relatively few records exist for wintering and migratory GCW [Golden-cheeked Warbler]. A thorough exploration of the known habitat types and other areas of similar habitat are needed. A remote sensing study and associated GIS that can be used to monitor the distribution and rate of change of suitable winter habitat for the GCW should be developed. The ground-truthing for this project should be coordinated with field survey activities called for in Task 1.35.

Not much of a contribution, you might say, considering all those meetings. However, I had argued that if $1.5 million a year could be spent on the golden-cheek's four-month stay in the Lone Star State, *something* should be spent trying to understand the rest of the bird's time during its annual cycle.

Carol and her bosses in USFWS Region II evidently took my suggestion seriously, because in June 1995, they provided me with the funding necessary to figure out just what the golden-cheek was up to once it left the Texas canyonlands.

You will recognize, dear reader, similarities in the above-cited paragraph in the recovery plan to our ongoing work in Costa Rica and Belize as described in earlier chapters. That work was underway while I was participating in the recovery team meetings, and I had argued there that these new technologies, coupled with fieldwork, might provide us with information on the relative value to an endangered migratory species of the different habitats occupied during the course of the annual cycle—a real-life test of the logical inference that those eight months spent away from Texas might have some importance.

It made sense. But the key word, as I knew well, was *might*. Compared with hundreds of records, detailed accounts, and even a book on golden cheek life in Texas, there were exactly twenty-nine published citations in the literature for the bird on its Central American wintering grounds, all of which were from the highlands of Honduras or Guatemala. Furthermore, these references consisted of little more than date and locality; there was nothing whatsoever about how the bird lived.

Figure 26-1. Professor Sherry (Pilar) Thorn, Marvin Martinez, and I in Golden-cheeked Warbler pine-oak wintering habitat near La Esperanza, Honduras.

I decided to begin my quixotic quest for wintering golden-cheeks in Honduras, based on the information in Burt Monroe's *Birds of Honduras*. His account was brief—seven specimens from only three localities—but Monroe's impeccable reputation for accuracy and thoroughness made it seem like the best place to start my investigation.

My next step was to find a Honduran ornithological professional who could familiarize me with the logistics and legalities of conducting fieldwork in the country and perhaps provide students willing to assist me with the research if they did not have the time or interest themselves. I turned to my colleagues in the NOS for help in locating this person and was quickly directed to Professor Sherry (Pilar) Thorn.

Sherry was famously the doyenne of Honduran ornithology—indeed, of Honduran biology. A couple of decades earlier, she had arrived in the country as a young, idealistic Peace Corps volunteer. She married a Honduran, had two children, and joined the Biology Department faculty of the University of Honduras, where she served as a full professor, instructing most of the Hondurans who went on to careers in medicine or teaching biology in the country. She also wrote a medical textbook for her students—in Spanish, of course. Much of what we were able to accomplish in Honduras was due to the knowledge and help she provided.

I found a university phone number for Sherry through the NOS membership listings and gave her a call in the summer of 1995 as I tried to develop a plan of operation for the work. I wasn't quite sure whether to open the conversation in Spanish or English but decided to start with Spanish, as that was the official language. Sherry, on answering, quickly realized that things would go more smoothly in English, so we switched to that tongue. She immediately expressed her willingness to help and insisted that I stay with her at her house in Tegucigalpa when I came to begin the project, which I planned to do in the fall.

I arranged to do a brief trip down to Honduras in October to get to know Sherry, work out the logistics of future visits, and see if I could find the bird. Accordingly, I flew out of Dulles Airport at 7:30 a.m. on October 17, got to Miami by 9:30, and left for Tegucigalpa at 11:30, arriving there at 11:30 local time (central standard time). I rented a little Datsun coupe and headed to Sherry's place. I spent some time that afternoon talking with her about the project and what I needed as far as logistics and permits were concerned. She had never seen the Golden-cheeked Warbler despite all of her years leading field trips for ornithological classes, which didn't sound encouraging. That evening, Dr. Becky Mynton and her daughter Ginny joined us for supper. Becky had a PhD in ecology from the University of Maryland. In fact, she was there at the same time as my boss at the Smithsonian, Chris Wemmer. She worked for the Honduran Ministry of the Environment and was able to clue me in on where to go to get topographic maps and satellite photos for the mountainous parts of the country, where our bird was most likely to be found.

The next day, the eighteenth, I spent scurrying around to various government and university offices to get the maps and permits needed. On the nineteenth, the great adventure began on which the entire success of the proposed multi-thousand-dollar, multiyear project hinged—namely, the attempt to locate an actual wintering Golden-cheeked Warbler. The following account of that critical event was written based on notes kept during the trip.

Wintering habitat for the golden-cheek, according to Monroe, was highland (above four thousand feet) pine-oak woodlands. The nearest locality to Tegu-cigalpa (twenty miles) where the bird had been found was a place with the euphonious name of "El Cantoral." On October 19, 1995, I took my beat-up Datsun and with Sherry's son, José, riding shotgun, headed up into the rug-ged mountains north of the sprawling capital. Once we had turned west off the main highway, the "road" to the tiny settlement was no more than a cart track, but we encountered no ruts that José couldn't push the little vehicle out of, and by one in the afternoon, I was walking through disturbed (burned, grazed) but decent pine-oak at about five thousand feet. Nonfield biologists may find my terror at this moment difficult to comprehend. For ten years I had been blis-tering Fish and Wildlife and the golden-cheek committee for doing nothing to learn about the eight months of the year when the bird was not in Texas. Finally, they had capitulated and given me a major grant to study the life history, population ecology, and distribution in its reputed winter range (mountains of Sierra Madre from southern Mexico to northern Nicaragua). Now, here I was, exposed and alone, figuratively poised over the bull's horns, two thousand miles from home in the classic "I can do that—how am I going to do that?" moment. Monroe had said the bird was "rare" in Honduras. He had reported only two localities for the whole country, and no one had seen it at El Cantoral in sixty years.

Most birds in this habitat at this time of year in the Central American Sierra associate in mixed-species flocks composed of one to a few individuals of eight to twenty different species. The first flock I found consisted of a Grace's Warbler,

Yellow-throated Warbler, Black-and-white Warbler, Painted Redstart, Eastern Bluebird, Western Wood-Pewee, Hepatic Tanager, and Chipping Sparrow, plus five Hermit Warblers and four Black-throated Green Warblers. These last two species are close relatives of the golden-cheek, and each Black-throated Green in particular had to be carefully examined because females of the two species can be quite similar. A half mile farther on, I found a second flock containing many of the same species seen in the first, plus the Unicolored Jay, Buff-breasted Flycatcher, and four Townsend's Warblers (another close relative of the golden-cheek). The third flock, located five hundred yards or so from the second, contained several Hermit, Black-throated Green, and Townsend's Warblers, a Painted Redstart, Blue-headed Vireo, Spot-crowned Woodcreeper, Wilson's Warbler, and . . . three male Golden-cheeked Warblers! Hallelujah!

So that's how it started. With advice from my dear friend and US Forest Service research scientist Dick DeGraaf, I hired Dave King to head up the fieldwork effort, assisted by several students of Sherry's when needed. Peter Leimgruber and Jeff Diez helped with the GIS and remote sensing.

The logistics of the fieldwork were quite different from any of my previous projects. Initially, using satellite remote-sensing data, we developed a map depicting the major habitats of the region by unique color back in our GIS lab in Virginia for habitats over three thousand feet in elevation constituting the known wintering range of the Golden-cheeked Warbler at the time. We then took this map into the field with us as we visited every highland site reachable by "road" (trails, mostly) along the entire six hundred miles of the Sierra Madre mountain spine, from Chiapas to southern Honduras, looking for the bird and tying the habitat in which it was found to a particular color.

Inns, or really any kind of lodging, were scarce, so we almost always had to camp. A typical outing would find us driving into a new location with apparently suitable golden-cheek habitat and setting up a six-person tent. We would then head out separately to survey for the bird for as long as the light held. *Surveying* meant strolling along listening for the call notes given by the noisier members of the kinds of flocks frequented by golden-cheeks—jays, tanagers, woodcreepers, and the like. The observer would then try to locate the flock and, if found, follow it until a thorough examination of the members either revealed golden-cheek presence or not. If a golden-cheek was found, the exact location was recorded using the GIS, and the habitat color on the map, the dominant trees of the habitat, and the species of the trees in which the bird was seen to forage were all identified. When the light failed, each person would head back to camp to enter the data into our logbooks. Then we would build a fire and fix supper, usually Dinty Moore beef stew supplemented with flour or corn tortillas. That done, we would write up our field notes and hit the sack. The next day began at 4:00 a.m., when we would rise, complete our ablutions, build a fire, and prepare breakfast, usually fried Spam and eggs, tortillas, and cowboy coffee. Thus fortified, we resumed surveys at first light, with each person carrying his own lunch, maybe a tortilla with PB&J (taco gringo) or a can of sardines.

We spent six months over two winter field seasons (December 1995–February 1996 and December 1996–February 1997) using these procedures in Chiapas, Guatemala, and Honduras.

What we found is that Golden-cheeked Warblers winter in highland (above four thousand feet) pine-oak habitat as individual members of mixed-species flocks from southern Chiapas to northern Nicaragua. This habitat is seriously threatened by clearing for timber, agriculture (coffee mainly), and pasture.

In addition to these illuminating data on the wintering biology of the Golden-cheeked Warbler, we were able to gain an "appreciation" for canned luncheon "meat" afforded to few, in honor of which Dave King compiled an authoritative compilation of Spam haiku, which I present below.

Spam Haiku

An anthology of the world's greatest poetic
tributes to the apotheosis of pork products

Authors Anonymous (for obvious reasons)
Compiled by the famed Neotropical ornithologist Dr. Dave King

1.
Oh . . . tin of pink meat
I ponder what Thou may be
Snout, ear, or feet?

2.
Anathema!
Proscribed by all that is Holy!
Thy evil pinkness beckons!

3.
In the cool of the morning
I fry up a sputtering slab.
In the distance . . . a dog barks.

4.
Pink tender morsel
Glistening with salty gel
What the Hell is it?

5.
Ears, snouts, innards
A homogeneous mass
Pass another slice!

6.
Cube of cold cerise
With yellow specks of porcine grease
Spork, please!

7.
Old man seeks doctor
"I eat Spam daily"
"Angioplasty!"

8.
Highly unnatural—
the tortured shape of this "food"
A small, roseate coffin

9.
Slicing your sweet self
Salivating in suspense
Sizzle, sizzle . . . Spam!

10.

Pink porcine temptress!
I can no longer remain
Vegetarian!

11.

Grotesque glowing mass
In blue can on shelf
Quivering . . .
Alone

12.

Pink, slimy, pumicy
My chunk of Spam
Shining sunlit on my plate

13.

Oh, Argentina!
Your little tin of meat
Soars o'er the Pampas!

14.

Ah!
Rosey fingered dawn!
A block of sunrise!
Spam!

15.

Little lump of meat
In pool of clear jelly
Now I heat the pan!

16.

Pork parts packed
In gelatinous goo

Shimmering!
Waiting!

17.
I think that I shall never see
A poem so lovely as art Thee
In Thy incandescent pinkness!

18.
Naked!
Pink!!
Stripped of Thy blue can!!!
Beckoning!

19.
The dogs bark.
The caravan passes on.
Spam remains.

At about the same time that our Central American work was being done, the USFWS funded a GIS/remote-sensing mapping effort for the entire Texas range. For reasons best known to the organization, these data were never published, but we got a copy of the final report from the contractor (David Diamond). Using this information in combination with our own, we were able to calculate the total amount and distribution of breeding and wintering habitat and to calculate the carrying capacity (total number of birds that could be supported) for these two parts of the species range.

The results, published in 2003 in the journal *Ecological Applications*, were stunning: the Texas breeding range had sufficient habitat to support an estimated 230,000 birds, whereas the Central American winter range had sufficient habitat to support only about 34,000 birds. These data contain a clear conservation message: population size for the Golden-cheeked Warbler is controlled by wintering-ground habitat availability.

This effort at carrying capacity estimation for both breeding and wintering ground in a migratory species is unique to my knowledge. Nevertheless, its import has broad implications for a number of other migrants whose populations are assumed to be threatened by breeding habitat loss (such as Kirtland's Warbler and Bobolink). It is necessary to examine the entire life cycle carefully before drawing conclusions or spending a lot of money on conservation programs likely to have little or no effect on species preservation.

There is a famous "perception as a requisite for reality" conundrum that goes as follows: "If a tree falls in a forest and nobody hears it, does it make a sound?" The paradox lies in the definition of the word *sound*. Obviously, the fall of the tree generates molecular movement, but such movement is sound only if there is someone around to perceive it and identify it as such.

I think that a similar paradox applies to scientific discovery as well. We spent $100,000 over five years of intensive study involving fieldwork and computer-assisted analysis of satellite imagery of Golden-cheeked Warbler wintering distribution, ecology, and habitat use throughout its range in the highlands of Middle America from Chiapas in the north to Nicaragua in the south. In addition, as described above, we incorporated satellite imagery analysis of breeding habitat distribution and likely population status in Texas to reach conclusions regarding where the greatest threat to the bird's population existed. The results were published in a series of papers in prominent ornithological and conservation journals in the late 1900s and early 2000s, yet you will not find this work referenced anywhere in summaries on the bird's life history or conservation status. Even the species account for the authoritative *Birds of the World*, the conclusion of which is quoted at the beginning of this chapter, mentions nothing whatsoever about these findings and places all conservation focus on the bird's brief time in Texas.

Why?

To answer this question, I must indulge in a bit of speculation. In my opinion, the reason for the suppression of our work relates directly to the falling tree conundrum. If publications containing inconvenient facts are ignored, then it is as though they do not exist. No one hears them. The facts in this case are that our data demonstrated unequivocally that destruction of highland pine-oak habitat on the wintering ground was the cause of golden-cheek decline. Now you may say "Well, isn't that finding a good thing? If you really want what is best for the bird, then knowing where the greatest threat lies should be hailed as a critically important step forward in its preservation, shouldn't it? After all, with this knowledge in hand, you can begin working to fix the real problem causing the bird's disappearance."

Or maybe not. Preservation is one thing. Conservation is another. Conservation is not science; it is policy with regard to public use of the environment, and in this case, Golden-cheeked Warbler conservation has served as the number-one tool limiting development in the Texas Hill Country. The battles in which this weapon has been used have been ferocious and continue to this day, with plenty of splenetic invective in the media from conservationists, developers, politicians, and many other interested parties. For the conservationists in this war to admit that the bird will need preservation in its wintering grounds in Chiapas, Guatemala, Honduras, and Nicaragua in addition to its breeding grounds might be a devastating blow to Texas conservation efforts. This problem, I think, is why our golden-cheek work has been buried. The "greater good" is satisfied, at least from one perspective. From another, such disdain for the scientific truth in service of a specific political end, however laudable it might seem, is deeply disturbing and saddening.

SHAZAM!

Migrants and Climate Change

Then felt I like some watcher of the skies
When a new planet swims into his ken;
Or like stout Cortez [Balboa, in fact] when with eagle eyes
He star'd at the Pacific—and all his men
Look'd at each other with a wild surmise—
Silent, upon a peak in Darien.

—JOHN KEATS, *ON LOOKING INTO CHAPMAN'S HOMER*

I was shuffling along on a "run" down the Welder ranch road bordering Pollito Lake on a brilliant day in late February 2006 when I heard an electrifying sound from a grove of shoreline cedar elms:

"Kiskadee!"

Now, there is only one thing in the world that makes that sound—namely, a Great Kiskadee, one of the best-known and earliest-described (1648, based on a bird shot in French Guiana) birds of the new-world tropics thanks to its raucous call, striking appearance, and open-country habits. Hearing one calling on that beautiful early spring morning in South Texas was a gobsmacking moment of the first order.

You may be forgiven, dear reader, if you fail to grasp the staggering potential of this discovery, which leaves me, like Keats in his poem quoted above, grasping for metaphors appropriate to convey to you the astonishing nature of this discovery.

To aid me in this endeavor, I will tell a little story. When I was a youth of nine or ten, I spent an inordinate amount of time searching for and attempting to capture the various vermin that lived near our house at 59 Chestnut Street in Jamestown, New York. It was during one of these outings that I found a horned toad.

Having watched the Disney program *The Living Desert* a few dozen times, I knew perfectly well what it was, and I also knew that such animals were not supposed to occur outside the hot sandy wastes of the Southwest. Nevertheless, here it was on a residential street in western New York state. I picked it up without incident (they depend mostly on camouflage to avoid capture) and took it home as a pet. Ants were supposed to be a favorite food according to the literature, but I could not get it to eat them or any other delicacy that I could provide, and it died of apparent sadness and/or inanition a couple of months later.

So where did it come from? I have always assumed that a human agent was involved—a neighbor of mine, perhaps, who had captured or purchased one on a trip to Arizona and let it loose when they found that its near complete lack of activity made it a disappointing pet. Doesn't matter, I guess. The point is that, like the kiskadee, I found it in a place where it should not be. However, unlike the kiskadee, there were perfectly logical explanations for its presence.

Kiskadees didn't breed at Welder, as I ought to know, since my secretary at the Connor Museum, Jan Ballard, had discovered a nest of one outside her office window in an ebony tree on the Texas A&I campus in Kingsville on May 21, 1982, the northernmost record for the species at the time and a long, long way south of Pollito Lake. Nevertheless, against all rules and regulations regarding range limits, the bird was here, and about two hours later, I was able to find its nest, a soccer ball–sized bundle of sticks and grasses ten feet up in a dead huisache out in the middle of the shallow oxbow lake, extending the known breeding range of the bird by sixty miles.

But how could I be so sure of the astounding character of this finding? The answer to this question comes in two parts, the first of which has to do with the remarkably detailed knowledge of Texas bird distribution.

When Texas joined the Union as the twenty-eighth state in 1845, it was known mainly for its size, remoteness, and the ferocity of its indigenous inhabitants. The Mexican War (1846–48) broadened and deepened that understanding by making several thousand soldiers from elsewhere across the country familiar with its awesome diversity of landscape, flora, and fauna, including James P. McCown, a West Point graduate from Tennessee who made extensive collections of birds from the lower Rio Grande Valley during his service in the region.

Based on Captain (later General) McCown's notes and specimens, two scientific papers were published in prominent journals of the time, and the secret was out: Texas has the richest birdlife of any region in the continent north of the tropics. Since then, the state's avifauna has received intense scrutiny from amateurs and professionals alike. By 1912, John K. Strecker, a professor at Baylor University, felt sufficiently confident of knowledge regarding the state's birds to publish a checklist that included concise information on their distribution. For the kiskadee, he states, "Lower Rio Grande (Cameron and Hidalgo counties), rare summer resident."

Harry Church Oberholser spent most of his impressive professional career focused on the topic. Dr. Oberholser first visited the state as a member of the US Biological Survey in 1900. By the mid-1920s, he was working to compile and summarize every piece of information that he could find on the abundance and distribution of all the bird species known from the state, with an eye toward publishing a book on the birds of Texas. He continued in this activity for the next forty years and was still working toward that end at the time of his death in 1963 at the age of ninety-three. Edgar Kincaid, who had been assisting Dr. Oberholser during the final period of this exhaustive endeavor, working with two associates, Suzanne Winkler and John Rowlett, published the work in two volumes in 1974.

The book contains maps depicting the distribution by season for every bird known from the state based on the tens of thousands of specimens and sightings contributed by thousands of collectors and observers. The work represents a thorough summary of what was known of Texas bird distribution in the early 1970s. Based on this information, the northernmost confirmed (by specimen) breeding record of the birds was from Cameron County along the lower Rio Grande River, 160 miles south of Welder. The northernmost record of suspected breeding (presumably based on a calling bird) came from Kenedy County, 90 miles south of Welder.

Now, you may ask (as many have) how could I *know* that these birds had not been breeding at Welder or other sites north of the known breeding range?

Well, as I have already explained to you, Dr. Oberholser's exhaustive tome said so. But I had much more detailed information than that on which to base my amazement. While Texas bird distribution in general is well known for reasons described above, that of the Coastal Bend region of the state, where Welder is located, is especially well documented due to its fame as the most diverse portion of a singularly diverse state. Several treatments have been published on the birds of the area, including one by myself and Gene Blacklock in 1985 (*Birds of the Texas Coastal Bend*, Texas A&M University Press), wherein we reported the Kingsville breeding record for the kiskadee mentioned above as the northernmost for the state.

But that isn't all. While bird distribution for the Coastal Bend is the best known for the state, the bird distribution of the Welder Refuge is the best known of any other part of that region. Since the refuge's establishment in 1954, its seven square miles have been the subject of intensive scrutiny by some of the most prominent experts on birds for the state and country, including its longtime director, Clarence Cottam, and longtime curator, the aforementioned Gene Blacklock. No evidence of breeding kiskadees was discovered during that half century of work. Furthermore, as described in earlier chapters, I conducted a bird-banding study at Welder in the mid-1970s, during which over fifty thousand net hours were accumulated, and over ten thousand birds over of ninety-five species were captured. No kiskadees were seen, heard, or captured by my team during the entire two years of that intensive study.

"OK, I get it! There weren't any kiskadees at Welder," you may say. "You can stop beating that dead horse."

Maybe so. But even a dead horse can bite you in the ass, as I was shortly to discover.

In any event, to my mind at least, breeding populations of the kiskadee had not been present at Welder, and now they were. But I knew that audiovisual records were worthless as documentation for findings as important as these. So I decided to test the reality of range change for this bird and any others for which the same phenomenon might have occurred.

With permission from Lynn Drawe, then director of the Welder Wildlife Foundation, I initiated a project to determine what new species of birds might be found at Welder and to chronicle whether the birds represented outliers or actual breeding populations. The plan was to begin in 2007 and continue through 2009, with a month or two spent at the ranch each spring during the breeding season attempting to answer the following questions:

1. Have any species of resident (present year-round) subtropical birds expanded their breeding ranges significantly (fifty miles or more) north of their range in Texas as known in 1974 (date of publication of *The Bird Life of Texas*) to include Welder and its immediate vicinity?

2. Have any species of migratory birds that breed in Texas and winter in the tropics expanded their breeding ranges significantly (fifty miles or more) north of their range in Texas as known in 1974 (date of publication of *The Bird Life of Texas*) to include Welder and its immediate vicinity?

3. Have arrival dates changed for migrants returning to breed at Welder that wintered in the tropics?

I began the study in March 2007 and did fieldwork through that spring as well as those of 2008 and 2009. My goal was to document the answers to the three questions listed above for the kiskadee and as many species in the categories mentioned in as many ways as I could think of, including

- recording the precise date and locality of paired (female present) singing males,
- recording the precise locality for as many active nests of these birds as I could find,
- recording the precise date and locality of adults feeding young, and
- recording the presence (date and number) of individuals of the species of concern at a feeding stationed monitored 24/7 by several thousand observers checking in via webcam over a three-year period (a collaboration with Professor Dezhen Song from Texas A&M University and some robotics folks from UC Berkeley).

Based on the data collected during that period, we were able to answer the three questions as follows:

1. Yes. Four species of resident subtropical birds have expanded their breeding ranges significantly (fifty miles or more) north of their range in Texas as known in 1974 to include Welder and its immediate vicinity: Great Kiskadee, Green Jay, White-tipped Dove, and Green Kingfisher.
2. Yes. Seven species of migratory birds that breed in Texas and winter in the tropics have expanded their breeding ranges significantly (fifty miles or more) north of their range in Texas as known in 1974 to include Welder or its immediate vicinity: Audubon's Oriole, Hooded Oriole, Groove-billed Ani, Cave Swallow, Tropical Parula, Buff-bellied Hummingbird, and Lesser Nighthawk.
3. Yes. Arrival dates have changed for at least three species of migrants returning to breed at Welder that wintered in the tropics: Summer Tanager, Brown-crested Flycatcher, and Yellow-billed Cuckoo. Members of these species arrive at least two weeks earlier now on average as compared with arrival dates from decades earlier.

Now we come to the promised second part of the Texas range change saga: What is it that makes these findings so astounding? To understand the answer to this question, you have to know something about current ideas regarding how the evolution of species works. Darwin suggested that evolution—that is, the gradual change in structure and function of organisms—works through the effects of the environment on small differences among individuals (natural selection). Successful differences are passed from parents to offspring, while those that are unsuccessful die out with their possessors. It took a while to figure out the mechanics of this operation, but eventually, it was found that structures called genes carried in the egg of the female and sperm of the male were the bearers of this information, which was passed on to their offspring, and the differences between genes were caused by mutation—that is, small changes in gene composition.

Based on this information, the great evolutionary biologist Ernst Mayr formulated the following proposition to explain the process of speciation—that is, the evolution of new species. Mayr stated that a species is made up of "groups of actually or potentially interbreeding natural populations reproductively isolated from all other such groups."

In other words, new species form by the process of random changes in genes when they are geologically separated from other members of their parent species—by distance or some other barrier, like a mountain range or an ocean.

This explanation has served well for evolutionary biologists trying to understand the process of how new species come about for at least the last half century or so since its formulation. There is just one problem with it—namely, the assumption that members of a given species do not wander around outside their known ranges. Indeed, the range—that combination of all ecological, environmental, and geographic factors that determine where a species can survive—is a genetically determined characteristic, which means that wanderers outside the range will not survive long enough to breed. That's the importance of the barrier. With the barrier in place, there is no possibility of genetic exchange between related populations, which allows each such isolated population the chance to go its own way, forming a new species.

Field biologists have always known that the assumption of genetic isolation of populations was wrong. Dispersers do it all the time. The problem has been that if this assumption is wrong, how can speciation through random mutation work? Since we do not know the answer to that question, the idea of frequent movement of dispersers outside their breeding range has been met with great resistance.

I knew all of this, of course, but I had not understood how impervious to facts the scientific establishment could be until I tried to publish the results of the Texas work, as I relate below.

By the close of our third field season in June 2009, I felt that we had enough information to publish a paper summarizing the results of our Welder research. I anticipated no difficulty in this endeavor. The data were perfectly straightforward: eleven species of subtropical birds had extended their breeding ranges sixty or more miles to the north in about thirty years. I had no solid explanation for why they had done this. I suggested climate change as a possibility, but really, it didn't matter what was the cause. The important finding was that rapid range change was a clear demonstration of the ubiquity and power of dispersal. If a new habitat suitable for breeding appeared, our data demonstrated unequivocally that dispersers would find it, and it wouldn't take them long.

I decided to submit the paper to *The Auk*, the most prominent bird journal at the time, confident that our findings would soon be placed before the world community of avian scientists. I was astonished, therefore, when I got an email from the editor telling me that he was not going to bother sending the paper out for review because it was certain to be rejected. Yikes! That was a shock. Not that I was a stranger to rejection. Dozens of my papers have been rejected, including several by *The Auk*. But why? On what basis could they reject the paper? The findings were clear and unequivocal: Thirty years ago, there were no breeding populations of these birds at Welder. Now they were there. The evidence was indisputable!

Well, no, the editor said. The evidence was not indisputable. In fact, he explained (as to a child), without a statistical test, it was not even evidence. That comment blew my mind. I was speechless, but only momentarily. I then said that he knew as well as I that zeros were useless in statistics. You have to have samples with some value other than zero in order to perform a comparison. Having such a requirement simply ignored a common occurrence in biogeography—namely, the presence of something where it had not been or the absence of something where it had been previously. Establishing this ridiculous condition was analogous to requiring a modern mammoth specimen in order to prove that the species disappeared from a place eight thousand years ago.

"All right," he said. "I'll send it out, but I know what will happen." And he was right. He gave it to an associate editor who simply rejected it out of hand for lack of a statistical analysis.

The infuriating issue here is that rejecting the paper had nothing to do with statistics. What the editors were saying is "We do not believe that there were no breeding individuals of these species at Welder thirty years ago. There had to be because ranges do not change in the matter of a few years. The obvious explanation for your findings is that the birds were there. Nobody looked hard enough to find them." Case closed.

So I gave the paper at a European symposium entitled Tropical Vertebrates in a Changing World, the proceedings of which were published in the monograph series of the Alexander Koenig Museum in Bonn in 2011. The Germans had no problem publishing our results, but I guess I would be surprised if more than a dozen people read it.

Of course, the entire episode is just water over the dam, now. The online crowdsource site for bird observations, eBird (https://ebird.org/home), established in 2010, has made the controversy as irrelevant as discussions of a terra-centric universe. With tens of thousands of people continuously reporting observations of bird sightings across the continent, dispersal outside the breeding zone has become visible for hundreds of species and rapid range change obvious.

The importance of this finding can hardly be overstated. Climate change has made entire regions of the world that were once unsuitable now habitable for breeding, a fact demonstrated by range change in hundreds of species in a matter of years.

Dire predictions have been made regarding the future of migratory birds as a result of climate change. The idea is that migrants are locked by their genes into their breeding and wintering ranges, the routes taken to travel between them, and their departure and return times. Elements of this idea are true, sort of. At least some parts are true for some species sometimes. However, our Texas data demonstrate that climate change is not necessarily a threat to migrant survival. So long as there is a suitable habitat reachable by dispersing individuals, they will find it and use it. And it will not take thousands, hundreds, or even tens of years to happen. As eBird has shown, you can watch it happen in real time thanks to the sharp eyes and reporting skills of thousands of dedicated reporters. We knew this even before eBird and our Texas work, of course. Climate change cycles of warming and cooling have occurred many times. The place where I am sitting in western New York has gone from ice sheets to cold steppe (with mammoths!) to mixed coniferous and deciduous forest every forty to one hundred thousand years for the past 2.5 million years, with the plant and animal communities appropriate to each newly created habitat tracking those changes with striking rapidity.

How is this possible? The quick answer is dispersal. So long as there is some place on the globe where their populations can persist, progeny of the survivors will find and repopulate any suitable environment created by the change.

"Well, then, how do you respond to the evidence documenting that billions of birds have already been lost as a result of climate change?" you might ask. To which I would say that I suppose you are alluding to the paper published in the journal *Science* by Ken Rosenberg, John Sauer, Pete Marra, and buddies showing population declines in a wide

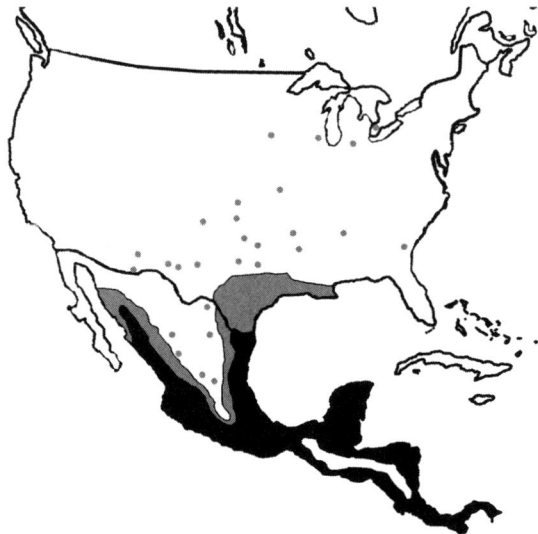

Figure 27-1. Range of the Great Kiskadee in 1974 (*black*) versus 2022 (*gray*). The 1974 range is based on data from Howell and Webb (1995) and Oberholser (1974); the 2022 range is based on data from eBird (2022). Note that a similar expansion has been observed into the temperate zone regions of Argentina in the South American portion of the species' range.

range of North American bird species over the past fifty years or so. To which you might answer "I read about it in the *New York Times*" or some similar media outlet. To which I would say "Yes, but the *Times* was simply reporting the data from Rosenberg et al. All such reports mentioning 'billions of birds lost' come from there."

So here is what I have to say. Population change in the Rosenberg paper was measured in two main ways: (1) radar and (2) the North American BBS (a coauthor of the Rosenberg study, John Sauer, heads up that organization). The BBS data are species specific. The radar data report the cumulative volume of migrants of various sizes by migration period. Both show large declines in migrants over the time periods for which they were measured (1970–2017). Nearly all of the declines reported have occurred in migratory bird populations that breed in eastern North America. Resident bird species actually have increased, and western species have not decreased. The authors hint darkly at a wide variety of causes, including climate change, requiring vast sums of money and effort to sort out for each individual declining species. However, in reality, these declines have nothing whatsoever to do with climate change. They are a direct result of habitat loss—breeding habitat for grassland, old field, and scrubby second-growth species and wintering habitat for forest-breeding migrants.

These same folks also claim that the declines come as a surprise. This claim is disingenuous at best. The declines are not a surprise. The ornithologist and diplomat William Vogt sponsored a symposium at the Smithsonian in 1966 bringing together the top tropical ornithologists in the world to consider the effects of tropical habitat loss on North American migratory birds. Theoretical biologists Steve Fretwell and John Terborgh predicted in the 1970s that the extensive clearing by subsistence farmers of Middle American tropical forest, where the majority of the migrants of eastern North America winter, would result in comparable declines in the populations of these species. My colleagues and I, as reported in earlier chapters in this book, documented that birds wintering in tropical forest needed those habitats for overwinter survival. We also summarized in several book chapters and other publications that the predicted declines were, in fact, occurring.

As for the sharp declines in grassland, old field, and shrubland species, that is an obvious result of the loss of their breeding habitats resulting from colonization in the early 1800s and the reversion of farmland to forest.

To summarize, then, total numbers of birds in North America are declining. They have been for at least the past fifty years for reasons given above—a fact documented in a large number of published, peer-reviewed studies. Despite the hysteria voiced by the media, there is no evidence to date that climate change is a major cause of population decline for any new-world bird species. In fact, as documented by our Texas work, many species of tropical and subtropical birds are increasing their population size by expanding their ranges into both Northern Hemisphere and Southern Hemisphere temperate regions.

Fine, you may say, but won't climate change eventually cause major negative changes in bird populations? The answer is yes. Of course it will. Just as it has at least nineteen times in the past 2.5 million years.

Up until seventeen thousand years ago, the entire northern half of the continent had no birds whatsoever—nothing but snow and ice. Based on what has happened before, we are currently nearing the peak of warming for this cycle. Cooling is likely to begin within the next few thousand years, a change that will once again result in the complete disappearance of approximately three hundred species of birds from what is now the temperate, boreal, and arctic regions of the continent. That ice age likely will last several tens of thousands of years, after which warming will commence once again. Or maybe not. What is important to remember, I think, is that climate change is a natural phenomenon, and while it may be true that human activities impact the rate of change, there is nothing that humanity can do to prevent it. So what happens to all these bird species when their ranges are under a mile or two of ice? Well, like the mammoth, the horse, and the camel in North America, if their habitats disappear completely, then they will disappear. However, for those whose habitats persist in the tropical and subtropical regions, they are likely to persist as well. And when warming returns, they will reoccupy northern regions again when forest and grassland replace the ice.

Climate change is caused by solar system factors that are not well understood. There is nothing whatsoever we can do about them except try to understand them better and plan as best we can for how to deal with their effects on our civilization.

CHAPTER 28

NEWTON, APPLES, AND
THE NEW PARADIGM

Anyone who interprets this
differently is mistaken.

—Qat'ada, quoted in Shihab al-Din al-Nuwayri,
The Ultimate Ambition in the Arts of Erudition

A sower went out to sow his seed: and as he
sowed, some fell by the way side; and it was
trodden down, and the fowls of the air devoured
it. And some fell upon a rock; and as soon as
it was sprung up, it withered away, because
it lacked moisture. And some fell among
thorns; and the thorns sprang up with it, and
choked it. And other fell on good ground, and
sprang up, and bare fruit an hundredfold.

—Luke 8:5–8

Evidently, the idea of gravity did not come to Sir Isaac Newton by being bonked on the head by an apple. That is just wrong. However, Newton avers that the inspiration *did* come to him by observing the fall of an apple in his father's orchard, confirming the importance of aha! moments in the formulation of important theories.

I have attempted to relate to the reader the several epiphanies I experienced that have served to punctuate the course of a long investigation into avian migration, the conclusion of which is that migration is a form of dispersal. That is the takeaway message from this work. All of the adventures regaled in the previous pages were steps along the path leading to this understanding. My advisor, Dwain Warner, and I first presented a form of this theory over forty years ago as a chapter in the second Smithsonian migration symposium volume. Since then, I have published over a hundred journal articles and three books on the subject, each including additional information in its support.

And what is the current consensus regarding migrant evolution? Honesty compels me to admit that it is *not* the dispersal theory. The northern home theory or its close theoretical relative, the migratory syndrome, in my opinion, form the most accepted paradigms. In both of these doctrines, migration evolves as a result of increasingly inclement weather forcing temporary seasonal abandonment of birds' breeding grounds. In fact, I have never seen any publication other than my own that presents or accepts a dispersal hypothesis for the origin of migration.

Given the abundance of evidence of many different kinds that I have presented herein, all of it rigorously peer reviewed for logic and accuracy, the reader must wonder why this should be so. I suggest two reasons. The first is that weather-related explanations for migration are simply "common sense." The second is that dispersal, like gravity, contagion, and continental drift, is invisible. All you can see as evidence of its existence is its effects—at least up until recently. Each of these ideas is examined a bit further below.

Migration as a Weather Phenomenon Is Common Sense

Recently, I read a novel written by a friend of mine, Dave (Oley) Olson, in which the lives of the various characters were presented as the product of their familial status and history and the circumstances into which they were born. Given these facts, their actions in response to the challenges and opportunities posed by their environments were not only predictable but inevitable. This idea derives from a philosophical doctrine known as determinism, which holds that all events, including the behaviors of humans and other animals, result from preexisting circumstances.

The beauty of this belief is that it is a tautology. *If* you had a mathematical value for the precise degree of influence that each aspect of the environment had on an individual of a given genetic makeup from the time of conception and *if* you had a formula that allowed you to accurately integrate and analyze all of these influences, then you could predict with precision how that individual would behave in response to each new event. Completely and irrefutably true. But these are big "ifs." As my maternal grandmother, Frisky Goodell, used to say, "If ifs and buts were candy and nuts, every day would be Christmas, John. So stop whining." The reality of determinism is that the possibilities involving only one or two such interactions likely are infinite, making predictions with certainty absurd.

Despite this little problem, most of us find determinism to be a useful tool in our everyday lives. We use it all the time to predict what will happen in response to this or that action. It only makes sense. Common sense. Nevertheless, we are all well aware that "common sense" is often wrong, but we just blow it off by saying, "Well, we really didn't have all of the key information." Which is true. And it is always true. And as soon as you admit that you don't have all of the information, you no longer have a tautology; you have a probability of unknown value. And by assuming that which you were trying to prove, your reasoning has become circular.

So what is my point? How does this stuff apply to the question at hand? The relevance of my little disquisition is that the northern home and migratory syndrome theories that argue that migration results from change in weather are deterministic. Both state that the genes of a bird born within a particular place, the range of its species, ensure that the bird will remain in that place unless some change in the environment forces it to move (migrate) in order to survive. Once the bird has left its range, it must persist through the period of inclement weather as a wanderer until the weather ameliorates, allowing its return. Stated in deterministic phraseology: obvious cause (weather) acting on known genetic makeup producing obvious effect (migration)—quod erat demonstrandum (QED).

Only because you have read this book, you now know that that is not the way migration works. Migration is not a deterministic phenomenon. It results from dispersal, and dispersal is a matter of free will. It's a choice. And no change in weather or any other aspect of the physical environment is required.

Dispersal Is Invisible

I have given lectures on the topic of dispersal many times over the course of my career, and I know from these experiences that the concept is a difficult one for most people to grasp. Even field biologists have some trouble with the principle, as illustrated by a central treatise of our discipline—namely, the biological species concept, which assumes dispersal to be limited by the species' known range. We now know that dispersal is not so limited. And how do we know this? Well, consider chapter 27 reporting our Texas work. We found that eighty species of birds had shifted their breeding ranges sixty miles or more to the north and east in the matter of a few years. No genetic change was required in order for this massive movement to occur. The only change needed was for environments formerly hostile to become suitable. Dispersers had always been passing through these environments. Nobody sent them any messages explaining that they should come on ahead.

Remarkably, many people still do not see how such range change is related to dispersal. They might see the logic, but they don't see the bird, I guess. It's still invisible. All we see is the result. For them we now have some additional proofs, thanks to technology. The first such proof comes from the phenomenal online data-logging and presentation system known as eBird. This system, operated by the Cornell Lab of Ornithology, provides a place where tens of thousands of volunteer observers can provide the date, time, and locality for their bird observations. The results are astounding. You can see migration as it occurs, day by day, for hundreds of species; you can see the core breeding range for common North American species; and you can see the many individuals documented to occur each year outside those ranges—the dispersers. Data loggers, tiny packs attached to birds and many other kinds of organisms, including butterflies and dragon flies, also demonstrate dispersal in action.

I think that a part of the problem that people have in understanding dispersal derives from its incomprehensibility. Why would an individual leave a place that has all the requisites for successful survival? Well, usually it's because their neighbors don't want them around anymore.

Finally, a fundamental problem for understanding migrant origins may derive from confused ideas regarding the concept of "home." *Home*, after all, often means "the place in which you were born." But it can also be used to describe any place where the

Figure 28-1. The quandary.

individual is able to find food and shelter for some variable period of time. In addition, the term can serve as shorthand for that portion of the life cycle from which migration movements began for any given population.

Early theorists for migration origins saw no problem with the use of the term *home* regardless of all its various meanings, since, in their view, they were all one and the same. The northern home was the place where the migrant was born, the place where it had its niche, and the place from which movement was initiated and subsequently evolved.

A half century of field research, as summarized briefly above, now demonstrates the problems with that view. The place where a migrant was born and raised need not be the place where the migrant's niche exists or from which migratory movement origi-nated. In fact, normally, they are quite different.

EPILOGUE

Farewell, farewell! but this I tell
To thee, thou Wedding-Guest!
He prayeth well, who loveth well
Both man and bird and beast.

He prayeth best, who loveth best
All things both great and small;
For the dear God who loveth us,
He made and loveth all.

The Mariner, whose eye is bright,
Whose beard with age is hoar,
Is gone: and now the Wedding-Guest
Turned from the bridegroom's door.

He went like one that hath been stunned,
And is of sense forlorn:
A sadder and a wiser man,
He rose the morrow morn.

—SAMUEL TAYLOR COLERIDGE,
THE RIME OF THE ANCIENT MARINER

So that's it. My tale is told, and you, dear reader, like the Wedding-Guest, are free to return to the present—not sadder, I hope, but certainly wiser. For now you know why it is that birds migrate: food and sex! And perhaps you will also have gotten the deeper message, the one alluded to in the final words of the Ancient Mariner and in my preface. We live in a world of surpassing beauty, wonder, and marvels, and while it is certainly our duty to care for it, we should not worry too much about its future. Huge changes are always a part of existence, but forever is a long time, and earth abides. Thousands of life-forms have appeared and disappeared, and it won't be long before our own species joins the paleo scene. Concern about that occurrence should not mar one moment of our brief time here.

SELECTED REFERENCES

Ackerman, J. 2020. *The Bird Way: A New Look at How Birds Talk, Work, Play, Parent, and Think*. New York: Penguin Press.

Brown, J. L. 1964. "The Evolution of Diversity in Avian Territorial Systems." *Wilson Bulletin* 6:160–69.

Buechner, H. K., and J. H. Buechner. 1970. "The Avifauna of Northern Latin America." *Smithsonian Contributions to Zoology*, no. 26, 1–119.

Dickerman, R. W. 1960. "Davian Behavior Complex in Ground Squirrels." *Journal of Mammalogy* 41:403.

Fisk, E. J. 1983. *The Peacocks of Baboquivari*. Tucson: Arizona Nature Conservancy.

———. 1985. *Parrots' Wood*. Manomet, MA: Manomet Bird Observatory.

Fretwell, S. D. 1972. *Populations in a Seasonal Environment*. Princeton, NJ: Princeton University Press.

Ladd, C., and L. Gass. 2020. "Golden-Cheeked Warbler (*Setophaga chrysoparia*), Version 1.0." In *Birds of the World*, edited by P. G. Rodewald. Ithaca, NY: Cornell Lab of Ornithology. https://doi.org/10.2173/bow.gchwar.01.

Mayr, E. 1963. *Animal Species and Evolution*. Cambridge, MA: Harvard University Press.

Parkes, K. C. 1963. "The Contribution of Museum Collections to Knowledge of the Living Bird." *Living Bird* 2:121–30.

Rappole, J. H. 1995. *The Ecology of Migrant Birds: A Neotropical Perspective*. Washington, DC: Smithsonian Institution Press.

———. 2013. *The Avian Migrant*. New York: Columbia University Press.

———. 2022. *Bird Migration: A New Understanding*. Baltimore, MD: Johns Hopkins University Press.

Rappole, J. H., D. I. King, and J. Diez. 2003. "Winter versus Breeding Habitat Limitation for an Endangered Avian Migrant." *Ecological Applications* 13:735–42.

Rappole, J. H., E. S. Morton, T. E. Lovejoy III, and J. Ruos. 1983. *Nearctic Avian Migrants in the Neotropics*. Washington, DC: US Fish and Wildlife Service.

Rappole, J. H., M. A. Ramos, and K. Winker. 1989. "Movements and Mortality in Wood Thrushes Wintering in Southern Veracruz." *Auk* 106:402–10.

Rappole, J. H., and D. W. Warner. 1976. "Relationships between Behavior, Physiology, and Weather in Avian Transients at a Migration Stopover Point." *Oecologia* 26:193–212.

———. 1980. "Ecological Aspects of Avian Migrant Behavior in Veracruz, Mexico." In *Migrant Birds in the Neotropics*, edited by A. Keast and E. S. Morton, 353–93. Washington, DC: Smithsonian Institution Press.

Rappole, J. H., D. W. Warner, and M. A. Ramos. 1977. "Territoriality and Population Structure in a Small Passerine Community." *American Midland Naturalist* 97:110–19.

Whitmore, R. C. 1981. "Review of *Migrant Birds in the Neotropics*." *Wilson Bulletin* 93:432–35.

Wiedensaul, S. 2021. *A World on the Wing: The Global Odyssey of Migratory Birds*. New York: W. W. Norton.

Willis, E. O. 1966. "The Role of Migrant Birds at Swarms of Army Ants." *Living Bird* 5:187–231.

———. 1975. *Sociobiology: The New Synthesis*. Cambridge, MA: Belknap Press.

Wynne-Edwards, V. C. 1962. *Animal Dispersion in Relation to Social Behavior*. Edinburgh: Oliver and Boyd.

INDEX

Page numbers in *italics* refer to figures.